SOIL QUALITY AND BIOFUEL PRODUCTION

Advances in Soil Science

Series Editors: Rattan Lal and B. A. Stewart

Published Titles

Advances in Soil Science

SOIL QUALITY AND BIOFUEL PRODUCTION

Edited by

Rattan Lal
B. A. Stewart

CRC Press
Taylor & Francis Group
Boca Raton   London   New York

CRC Press is an imprint of the
Taylor & Francis Group, an **informa** business

CRC Press
Taylor & Francis Group
6000 Broken Sound Parkway NW, Suite 300
Boca Raton, FL 33487-2742

First issued in paperback 2017

© 2010 by Taylor and Francis Group, LLC
CRC Press is an imprint of Taylor & Francis Group, an Informa business

No claim to original U.S. Government works

ISBN 13: 978-1-138-11783-9 (pbk)
ISBN 13: 978-1-4398-0073-7 (hbk)

Library of Congress Cataloging-in-Publication Data

Soil quality and biofuel production / editors: Rattan Lal, B.A. Stewart.
      p. cm. -- (Advances in soil science)
    Includes bibliographical references and index.
    ISBN 978-1-4398-0073-7 (hardcover : alk. paper)
      1. Biomass energy--Environmental aspects. 2. Soils--Quality. I. Lal, R. II. Stewart, B. A. (Bobby Alton), 1932- III. Series: Advances in soil science (Boca Raton, Fla.)

TP339.C367 2010
662'.6692--dc22   2009042283

Visit the Taylor & Francis Web site at
http://www.taylorandfrancis.com

and the CRC Press Web site at
http://www.crcpress.com

Contents

Preface

Traditional biofuels (e.g., wood, crop residues, animal dung) have been used as sources of household energy (such as for cooking and heating) since the dawn of human civilization. Modern biofuels go back to 1896 when Henry Ford's first car was designed to run on ethanol. They achieved prominence during the oil embargo of the 1970s. However, the current interest in biofuels is driven by high energy demands and the impacts of fossil fuel combustion on emissions of CO_2, CH_4, and N_2O with the attendant changes in climate. Global energy consumption is expected to grow by 50% between 2005 and 2030. From its humble beginning in the late 19th and early 20th centuries, fuel ethanol production in the United States grew to 310 million liters (ML) in 1980, 2.3 billion liters (BL) in 1985, 2.8 BL in 1990, 4.9 BL in 1995, 6.3 BL in 2000, 15.4 BL in 2005, and 25.9 BL in 2007.

Of the current annual global production of 700 million tonnes (Mt) of corn grains, 9% (63 Mt) are used for ethanol production. Production of ethanol from corn grains and biodiesel from soybeans has been blamed for high food prices (food-related riots reported from 30 countries in 2008), and conversion of forests and prairies to new land for crop production, leading to more anthropogenic emissions of greenhouse gases. The competition for food based on the needs of more than 1 billion food-insecure people that continues to increase because of high food prices has created an urgency to develop technology for producing cellulosic ethanol.

One potential source of ligno-cellulosic biomass is harvesting crop residues, especially those of corn, wheat, barley, rye, etc. This means the agricultural industry is now required to produce huge amounts of biomass along with grains for food and fuel, and this increased demand clearly impacts soil and environment quality. Indeed, indiscriminate removal of crop residues may exert severe adverse impacts on soil quality especially because of increased susceptibility to crusting and compaction, declines in soil organic matter, and accelerated water runoff and soil erosion.

Continuous removal of all or most of the above-ground biomass is simply not a sustainable system. More and more nutrient inputs will be required over time, and accelerated soil erosion will be exacerbated by water on sloping lands and by wind on flat lands. Some ecologists have suggested harvesting of prairie grasses (low intensity, high diversity biomass). Without replacing plant nutrients harvested in the biomass, it will not be possible to sustain repeated harvests of two or three cuttings per year. Another suggestion is to produce biomass by establishing energy plantations on degraded or desertified soils. Energy plantations can sustain high biomass production if they are established on good soils and sustained by effective practices of soil, water, and nutrient management. Expecting high and sustained biomass production without input on degraded soils is a myth at best, and a delusion at worst.

In addition to these environmental issues, some argue that the energy contained in biofuels (grain-derived ethanol) is insufficient to be economically or environmentally viable. Rather than an increase of 10% to 20% in energy, the ratio of energy output to energy input in biofuel production must be large (as with ethanol production from sugarcane in Brazil). Because of these biophysical, economic, environmental, and social factors, biofuels have often been cynically called "biofoolery" and "pie in the sky."

The strong interactions of energy, climate, food, and soil quality cannot be ignored. While ethanol made from biomass (rather than grains) provides unique economic and environmental benefits, the logistics of producing biomass in a sustainable manner remains a major challenge. Realizing the biofuel vision would require 1 billion tonnes of dry cellulosic biomass in the United States and about five times in the rest of the world. Assuming an optimistic biomass yield of 10 tonnes/ha of dry matter under high productivity conditions, an additional 100 million ha (Mha) of good quality land would be required in the United States and 500 Mha elsewhere in the world. Further assuming that average nutrient contents in cereal residues are 1% N, 0.1% P, and 1.5% K, additional annual removal rates would be 10 Mt N, 1 Mt P, and 15 Mt K in the United States compared with 50 Mt N, 5 Mt P, and 75 Mt K throughout the rest of the world, much of which would have to be replaced with fertilizers.

Production, packaging, distribution, and application of fertilizers also have hidden C costs, equivalent to 65 Mt (Tg) of CO_2-C for N, 1 Tg for P, and 12 Tg for K on a global scale. Not replacing the nutrients harvested in biomass would severely jeopardize soil quality by depleting organic matter and nutrient pools. In addition to the need for land area and nutrients, energy plantations would also require water for plant growth and biomass production. Water is a scarce and precious commodity required also for conversion of biomass into liquid biofuels. Water will become an increasingly limiting factor in a warming climate on an overcrowded planet.

For biofuels to become realities and to achieve production targets by 2030 and beyond, the importance of land, water, and nutrient requirements for producing biofuel feedstocks cannot be overemphasized. After the biophysical necessities (land, water, nutrients) are identified, soil and site agronomic research will be needed to develop best management practices (BMPs) for the dedicated species (switchgrass, miscanthus, elephant grass, poplar, acacia). As with agronomic and horticultural crops, the goal of developing BMPs for energy plantations includes enhancement of and sustaining of high productivity while maintaining and improving soil quality.

Therefore, this volume in the Advances in Soil Science series focuses on the issues of soil quality and biofuel production. The next volume will specifically address the issue of competition between food and biofuel. The need for water for food and fuel production may also be discussed in subsequent volumes.

Rattan Lal
B.A. Stewart

Editors

Rattan Lal, PhD, is a professor of soil physics in the School of Natural Resources and director of the Carbon Management and Sequestration Center, FAES/OARDC at Ohio State University. Prior to joining Ohio State in 1987, he was a soil physicist at the International Institute of Tropical Agriculture, Ibadan, Nigeria for 18 years. In Africa, Dr. Lal conducted long-term experiments on land use, watershed management, soil erosion processes as influenced by rainfall characteristics, soil properties, methods of deforestation, soil-tillage and crop-residue management, cropping systems including cover crops and agroforestry, and mixed/relay cropping methods. He also assessed the impact of soil erosion on crop yield and related erosion-induced changes in soil properties to crop growth and yield. Since joining Ohio State University, he has continued research on erosion-induced changes in soil quality and developed a new project on soils and climate change. He has demonstrated that accelerated soil erosion is a major factor affecting emission of carbon from the soil to the atmosphere and that soil-erosion control and adoption of conservation-effective measures can lead to carbon sequestration and mitigation of the greenhouse effect. Other research interests include soil compaction, conservation tillage, mine soil reclamation, water table management, and sustainable use of soil and water resources of the tropics for enhancing food security. Dr. Lal is a fellow of the Soil Science Society of America, American Society of Agronomy, Third World Academy of Sciences, American Association for the Advancement of Sciences, Soil and Water Conservation Society, and Indian Academy of Agricultural Sciences. He is the recipient of the International Soil Science Award of the American Society of Agronomy, the Hugh Hammond Bennett Award of the Soil and Water Conservation Society, and the Borlaug Award as well as an honorary Doctor of Science degree from Punjab Agricultural University, India and the Norwegian University of Life Sciences, Aas, Norway. He is past president of the World Association of Soil and Water Conservation and the International Soil Tillage Research Organization. He is a member of the U.S. National Committee on Soil Science of the National Academy of Sciences (1998 to 2002, 2007-). He served on the panel of the National Academy of Sciences on Sustainable Agriculture and the Environment in the Humid Tropics and has authored and coauthored approximately 1,375 research publications. Additionally, he has written 13 books and edited or coedited 43 other books.

B.A. Stewart, PhD, is distinguished professor of Soil Science and West Texas A&M University, Canyon Texas. He is director of the Dryland Agriculture Institute and former director of the USDA Conservation and Production Laboratory at Bushland, Texas; past president of the Soil Science Society of America; and member of the 1990–1993 Committee on Long-Range Soil and Water Policy, National Research Council, National Academy of Sciences. He is a Fellow on the Soil Science Society of America, American Society of Agronomy, Soil and Water Conservation Society, a recipient of the USDA Superior Service Award, a recipient of the Hugh Hammond

Bennett Award of the Soil and Water Conservation Society, and an honorary member of the International Union of Soil Sciences in 2008. Dr. Stewart is very supportive of education and research on dryland agriculture. The B.A. and Jane Anne Stewart Dryland Agriculture Scholarship Fund was established in West Texas A&M University in 1994 to provide scholarships for undergraduate and graduate students with a demonstrated interest in dryland agriculture.

Contributors

M. Bernoux
Institut de Recherche pour le
  Dévelopement
Paris, France

W.E.H. Blum
University of Natural Resources and
  Applied Life Sciences
Vienna, Austria

Karina Cenciani
Universidade de São Paulo
São Paulo, Brazil

Carlos Clemente Cerri
Centro de Energia Nuclear na
  Agricultura
Universidade de São Paulo
Piracicaba, Brazil

Carlos Eduardo P. Cerri
Escola Superior de Agricultura Luiz de
  Queiroz
Universidade de São Paulo
Piracicaba, Brazil

Marília Barbosa Chiavegato
Universidade de São Paulo
São Paulo, Brazil

M.J. Cruse
Department of Agronomy
Iowa State University
Ames, Iowa, USA

Richard M. Cruse
Department of Agronomy
Iowa State University
Ames, Iowa, USA

Brigitte Josefine Feigl
Universidade de São Paulo
São Paulo, Brazil

Leidivan Almeida Frazlão
Universidade de São Paulo
São Paulo, Brazil

M.V. Galdos
Centro de Energia Nuclear na
  Agricultura
Universidade de São Paulo
São Paulo, Brazil

M.H. Gerzabek
University of Natural Resources and
  Applied Life Sciences
Vienna, Austria

K. Hackländer
University of Natural Resources and
  Applied Life Sciences
Vienna, Austria

R. Horn
Institute for Plant Nutrition and Soil
  Science
Christian-Albrechts-Universtät
Kiel, Germany

Jane M.F. Johnson
U.S. Department of Agriculture
Agricultural Research Service
Morris, Minnesota, USA

Maysoon M. Mikha
U.S. Department of Agriculture
Agricultural Research Service
Akron, Colorado, USA

Sharon K. Papiernik
U.S. Department of Agriculture
Agricultural Research Service
Morris, Minnesota, USA

D. Pimentel
Cornell University
Ithaca, New York, USA

Jozsef Popp
Hungarian Academy of Sciences
Research Institute for Agricultural
    Economics
Budapest, Hungary

D.C. Reicosky
U.S. Department of Agriculture
Agriculture Research Service
Morris, Minnesota, USA

F. Reimoser
University of Veterinary Medicine
Vienna, Austria

Kurt A. Spokas
U.S. Department of Agriculture
Agricultural Research Service
St. Paul, Minnesota, USA

Mark D. Tomer
U.S. Department of Agriculture
Agricultural Research Service
Ames, Iowa, USA

Sharon L. Weyers
U.S. Department of Agriculture
Agricultural Research Service
Morris, Minnesota, USA

W. Winiwarter
Austrian Research Centers
Vienna, Austria
and
International Institute for Applied
    Systems Analysis
Laxenburg, Austria

S. Zechmeister-Boltenstern
Research and Training Centre for
    Forests, Natural Hazards, and
    Landscape
Vienna, Austria

F. Zehetner
University of Natural Resources and
    Applied Life Sciences
Vienna, Austria

1 Soil Processes and Residue Harvest Management

Jane M.F. Johnson, Sharon K. Papiernik,
Maysoon M. Mikha, Kurt A. Spokas,
Mark D. Tomer, and Sharon L. Weyers

CONTENTS

INTRODUCTION

The United States is interested in expanding renewable energy resources to address the interrelated problems of finite fossil fuels and global climate change by changing the energy paradigm from one based almost solely on fossil fuels to another that integrates multiple renewable energy platforms (Johnson et al., 2007d). Plant biomass feedstocks will be among the sources of renewable energy. Ethanol from corn (maize; *Zea mays.* L.), grain, and sugarcane (*Saccharum officinarum* L.) and biodiesel fuel from soybeans (*Glycine max* L.) and other oilseed crops are already used for transportation fuels. However, alone they are insufficient to replace petroleum (Perlack et al., 2005). Interest in using non-grain, cellulosic biomass has increased recently (Perlack et al., 2005). Agricultural and forest products represent potential non-grain biomass feedstocks for thermochemical (pyrolysis and gasification) and sugar (fermentation) platforms. Thermochemical technologies can substitute biomass for natural gas or coal and can also be used for producing liquid (pyrolysis oil) and solid (biochar) fuels (Islam and Ani, 2000; Gercel, 2002; Yaman, 2004).

1

Potential cellulosic biomass feedstocks are numerous and include woody and herbaceous perennial species, lumber industry wastes, forage crops, industrial and municipal wastes, animal manure, crop residues, and agricultural wastes or co-products such as bagasses and cannery wastes (FAO, 2004; Perlack et al., 2005; Johnson et al., 2007d). Corn stover and wheat (*Triticum aestivum* L.) straw are grown in sufficient quantities to support commercial-sized cellulosic ethanol production (Dipardo, 2000; Hettenhaus et al., 2000; Nelson, 2002; Graham et al., 2007). The sources and importance of individual feedstocks vary with location. Regional sources such as sugarcane bagasse and rice (*Oryza sativa* L.) may individually make only local contributions, but collectively they significantly help satisfy United States energy needs (Dipardo, 2000).

Several demands compete for non-grain crop biomass (a term used interchangeably with *crop residue* in this chapter). Small grain straw and corn stover are used for animal bedding and high-fiber feed. Burning corn cobs and other cellulosic materials for heating or cooking was a relatively common practice less than a century ago, and still occurs in some locations. Straw is considered viable as a low-cost building or insulation material (Bainbridge, 1986; Yang et al., 2003). From a soil perspective, keeping non-grain biomass in the field returns essential nutrients for subsequent crops, maintains soil organic matter (SOM), promotes soil aggregate stability, and provides groundcover to reduce erosion (Johnson et al., 2006a).

Most estimates of the amounts of crop residues available for harvest are based on the sole constraint of minimized soil erosion (Lindstrom, 1986; Nelson, 2002; Perlack et al., 2005; Graham et al., 2007). Soil loss tolerance (T) was defined in 1997 as the average annual erosion rate (mass per area per year) that can occur and still permit a high level of crop productivity to be sustained economically and indefinitely by the United States Department of Agriculture (USDA) Natural Resource Conservation Service (NRCS). Nelson (2002) completed a three-year (1995–1997) county-level evaluation of residue removal rates that would provide soil erosion rates less than T. This analysis suggests that an average of 43 million Mg of corn stover and 8 million Mg of wheat could be removed annually for biofuel production (Nelson, 2002). Graham et al. (2007) estimated that sufficient stover was available in central Illinois, northern Iowa, southern Minnesota, and along the Platte River to support large biorefineries. Harvest rates were limited to amounts that maintained erosion rates less than T, and the study assumed all lands included were managed without tillage (Graham et al., 2007). However, Wilhelm et al. (2007) noted that residue requirements for maintaining soil organic carbon (SOC) exceeded those needed to limit erosion at or below T for a similar geographic area. The assessments conducted by Nelson (2002) and Graham et al. (2007) constrained harvest rates only to limit erosion; however, they provide a basis for more detailed analyses including the impact of residue removal on C cycling, future crop productivity, and other important considerations raised by Wilhelm et al. (2004; 2007).

Harvest of non-grain biomass has the potential to directly and indirectly affect many soil physical, chemical, and biological processes. Similar issues are raised concerning soil quality and sustainability for all proposed bioenergy platforms. Understanding the impacts of non-grain biomass harvest on soil processes will aid in developing harvest management systems including utilization of by-products to offset harvest impacts, thus optimizing potential benefits and reducing risks. Harvest guidelines and BMPs are necessary to protect soil from degradation. In this chapter, we discuss soil processes impacted by residue management with emphasis on non-grain biomass harvest. Soil processes reviewed include temperature, moisture and energy balance, C cycling, soil biology, nutrient cycling, soil aggregation, soil erosion, and watershed hydrology.

TEMPERATURE, MOISTURE, AND ENERGY BALANCE

The impacts of residue management on temperature (McCalla, 1943) and moisture (Duley and Russel, 1939) have been researched for more than 60 years, especially in the context of tillage. Surface residues modify the soil microclimate (moisture and temperature) primarily by altering the surface energy balance (Enz et al., 1988; Steiner, 1994; Horton et al., 1996; Wilhelm et al., 2004). Specifically, residue adds a boundary layer between the soil and atmosphere (Enz et al., 1988) that changes the corresponding energy inputs into the soil system (Horton et al., 1996). The net radiation for a bare soil is represented by:

$$R_{net} = S_{sky} - \alpha(S_{sky}) + L_{sky} - L_{soil} \qquad (1.1)$$

where R_{net} is the net radiation; S_{sky} is the incident short-wave (solar) radiation; α is the surface albedo (fraction of radiation reflected from the surface); L_{sky} is the incident long-wave sky radiation; and L_{soil} is the emitted long-wave radiation from the soil (Horton et al., 1996; Hillel, 1998). As indicated by Hillel (1998), day and night energy balances exhibit a major difference (Figure 1.1). At night, S_{sky} is negligible and the soil long-wave radiation is typically larger than the long-wave sky radiation, resulting in a negative net radiation flux at night and as a result, net energy movement is from the soil to the atmosphere. For a bare soil, the net radiation is the difference between the energy absorbed and lost by the soil. The net radiation on a bare soil can be apportioned as (1) sensible heat, (2) energy to heat the soil, or (3) energy used to evaporate soil moisture (Figure 1.1). However, the addition of a residue layer provides additional sources and sinks of energy (Ross et al., 1985; Bristow et al., 1986; Chung and Horton, 1987; Enz et al., 1988).

In addition to the processes described for a bare soil, the residue layer can (1) reflect more or less of the incoming radiation, depending on the residue surface albedo (Table 1.1); (2) utilize some of the incoming radiation to heat the residue layer; (3) use energy to evaporate water from the residue; (4) add increased resistance to water vapor fluxes from the soil, thereby reducing soil evaporation flux, and (5) transmit remaining energy to the soil surface (Shen and Tanner, 1990; Horton et al., 1996). Typically residues are lighter in color than soils, thereby increasing the albedo of a residue-covered surface compared to bare soil (Sharratt and Campbell, 1994; Table 1.1). Residues also trap a significant amount of air within the residue layer, thus significantly reducing the effective thermal conductivity of the material layer and reducing the amount of heat transmitted through the residue (Pratt, 1969). Thus, the amount of energy incoming to the soil surface will be less with a residue layer present compared to bare soil. Residues can intercept 50% to 80% of incoming radiation

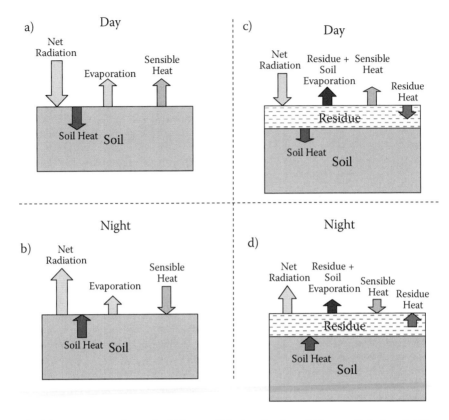

FIGURE 1.1 Energy fluxes for (a) bare soil during daytime, (b) bare soil during nighttime, (c) residue-covered soil during daytime, and (d) residue-covered soil during nighttime conditions. Arrows indicate directions of energy movement. Relative sizes of arrows approximate one potential scenario of dynamics of energy fluxes. Residue cover reduces energy gained (during daytime) and lost (at night) by dry soil surface. (*Sources:* Hillel, D. 1998. *Environmental Soil Physics*. Academic Press, San Diego, CA; Horton, R., K.L. Bristow, G.J. Kluitenberg, and T.J. Sauder. 1996. *Theor. Appl. Clim.* 54:27–37.)

(S_{sky}), keeping the surface soil temperatures within 20°C of ambient, whereas bare soil temperatures may rise 30°C or more above ambient (Ross et al., 1985).

The residue layer also impacts the aerodynamic boundary layer conditions of the soil surface (van Bavel and Hillel, 1976; Hagen, 1996). Residues typically increase surface roughness and correspondingly impact surface exchanges of heat and water and reduce soil loss by wind erosion. Residues increase infiltration and decrease evaporation, generally resulting in a net increase in soil moisture (Smika and Unger, 1986; Blevins and Frye, 1993; Wells et al., 2003; Govaerts et al., 2007a). In regions that experience significant amounts of wind-blown snow, surface residues trap snow, reducing frost penetration depth due to the insulating properties of the snow pack (Benoit et al., 1986). In addition, the snow surface is typically smooth due to low winter evaporation, producing additional soil moisture in the spring (Sauer et al.,

TABLE 1.1

Albedo Comparisons for Several Crop Residues and Materials

Material	Albedo*	Citations
Bare soil	0.04 to 0.40	Lobell and Asner, 2002; Markvart and Castañer, 2003
Green grass	0.25	Markvart and Castañer, 2003
Growing crops	0.10 to 0.40	Stanhill et al., 1966; Al-Yemeni and Grace, 1995
Maize (*Zea Mays* L.) residue	0.31 to 0.46	Tanner and Shen, 1990
Barley (*Hordeum vulgare* L.) straw	0.42 to 0.50	Novak et al., 2000
Sugarcane (*Saccharum officinarum* L.) residue	0.31	Bussière and Cellier, 1994
Wheat (*Triticum aestivum* L.) Straw	0.48 to 0.70	Major et al., 1990
Snow	0.70 to 0.90	Markvart and Castañer, 2003

* Albedo values depend on residue moisture content; typically, the higher the moisture content, the lower the albedo.

1998). Furthermore, trapped snow provides additional soil water recharge during spring thaws (Benoit et al., 1986).

As shown in Figure 1.1, residues impact the surface energy balance by reducing diurnal energy gain and loss at the soil surface. The resulting changes in soil temperature and moisture are functions of the physical properties of the residues and the conditions of the soil (Bristow et al., 1986; Steiner, 1994). In general, most studies agree that with increased residue cover (i.e., decreased crop residue removal), soil moisture content is increased (Russel, 1940); soil temperature maximums decrease, and minimums increase (Blanco-Canqui et al., 2006a). Consequences of these temperature and moisture impacts from residue coverage are less clear. All these factors depend on the interactions of altered soil microclimate conditions with other factors (soil type, climate, and crop type). Crop emergence has been shown to be sensitive to alterations in soil microclimate (Ford and Hicks, 1992; Drury et al., 2003). These resulting effects can be beneficial (Linden et al., 2000; Dam et al., 2005; Blanco-Canqui et al., 2006a), detrimental (Munawar et al., 1990; Liu et al., 2004), or negligible (Bristow, 1988; Swan et al., 1994) for crop emergence, development, and yield. Lower yields observed with high residue covers are hypothesized to result from slower soil warming during seed germination, lower pH, nutrient immobilization, and higher incidence of weeds and pests under residues (Cox et al., 1990; Mann et al., 2002; Drury et al., 2003; Jiang and Thelen, 2004; Liu et al., 2004). Delayed soil warming may delay planting, thereby offsetting gains of soil moisture retention (Nafziger et al., 1991). However, in drought-stressed areas, increased soil moisture can be vital (Power et al., 1986; Power et al., 1998; Jalota et al., 2001; Kato et al., 2007).

Surface residues insulate the soil surface, reducing diurnal temperature fluctuations in a residue-covered soil compared to a bare soil (Buerkert, 2000). In Minnesota, tall (0.6 m) corn stubble reduced frost penetration by 0.5 m and increased the minimum soil temperature by 2°C compared to soil with no residue, leading to a 25-day decrease to begin spring thaw (Sharratt, 2002). However, these beneficial properties of reducing frost depth and increasing soil moisture (Sharratt et al., 1998) resulted in lower spring soil temperatures. This is problematic because these factors may delay spring field operations and significantly impede early germination (Liu et al., 2004). Crop residue impacts on soil microclimate affect both crop emergence and growth and also the timing of the biological production of N_2O (Wagner-Riddle et al., 2008), weed pressure (Garcia-Huidobro et al., 1982; Shafii and Price, 2001; Duppong et al., 2004; Dhima et al., 2006), and C and nutrient cycling (Bayer et al., 2006).

SOIL ORGANIC MATTER (SOM)

Many physical, chemical, and biological characteristics of high quality soils are related to SOM (Doran and Parkin, 1994). Soil biota, nutrient cycling, residue decomposition, humification, and SOM cycling are interrelated. Soils tend to be more productive when organic matter is added regularly and allowed to decompose, thus stimulating nutrient and C cycling and maintaining or enhancing soil structure (Albright, 1938; Kumar and Goh, 2000). Soil organic matter enhances aeration, permeability, water retention, cation exchange, and buffer capacity (Stevenson, 1994; Kumar and Goh, 2000) and reduces soil compactability (Guérif, 1990; Soane, 1990; Diaz-Zorita and Grosso, 2000; Krzic et al., 2004).

Soil compactability is likely to increase if biomass harvest lowers SOM. Within one year, soil bulk density in the surface 6 cm was related inversely to the amount of corn stover returned on silt loam and clay loam soils in Ohio (Blanco-Canqui et al., 2006b). The highest bulk density was 1.45 Mg m^{-3} when all harvestable stover was removed compared to 1.24 Mg m^{-3} with 10 Mg ha^{-1} stover returned. Similarly, cone index and shear strength measurements also decreased with increasing surface stover.

C CYCLING

Plant roots and unharvested above-ground biomass provide the raw materials for building SOM. Photosynthate (organic C) enters the below-ground food web and traverses through multiple trophic levels before returning to the atmosphere with only a small fraction humified into stable SOM. As reviewed by Wilhelm et al. (2004), the amount of plant residue C in soil decreases over time through decomposition; within two years, less than 20% remains in the soil. These authors suggested that the small amount of new C converted to stable SOM implied that a large biomass influx was needed to provide substrate in excess of the respiratory demand of soil fauna. Soil organic matter is about 56% organic C (Stevenson, 1994). Soil organic C (SOC) is frequently used as a proxy to estimate SOM. A simple one-component model of SOC turnover using first-order kinetics (Equation 1.2) is useful where input for more complex models is lacking (Bayer et al., 2006):

$$C_t = C_0 e^{-k_2 t} + \frac{k_1 A}{k_2}(1 - e^{-k_2 t})$$

$$(1.2)$$

where C_t is the SOC stock at time t; C_0 is the initial SOC at time $t = 0$; k_2 is the annual rate of SOC loss by mineralization and erosion; k_1 is the annual rate of added C humi-fied into SOM; and A is annual rate of C addition. The first derivative of Equation 1.2 can be expressed as Equation 1.3 (Bayer et al., 2006; Huggins et al., 2007):

$$\frac{dC}{dt} = k_1 A - k_2 C \tag{1.3}$$

Simply stated, the change in SOC over time is a function of the rate of humification minus the rate of mineralization (inputs minus outputs). At equilibrium, dC/dt goes to zero and $k_1 A = k_2 C$ and SOC content reaches dynamic equilibrium C_e, as noted by Bayer et al. (2006):

$$C_e = \frac{k_1 A}{k_2} \tag{1.4}$$

Conversely, it is possible to use this simple model to solve for the annual rate of C addition (Huggins et al., 2007) at C_e:

$$A = \frac{k_2 C_e}{k_1} \tag{1.5}$$

If $k_1 A$ exceeds $k_2 C$, then SOC should increase; if not, SOC will decrease. If k_1 and k_2 remain constant, the change in soil C is proportional to inputs for a given manage-ment system. Several studies indicate that the amounts of both SOC and C inputs increased linearly (Larson et al., 1972; Paustian et al., 1997; Wilhelm et al., 2004; Follett et al., 2005; Kong et al., 2005; Bayer et al., 2006; Johnson et al., 2006a). However, other results reveal that SOC sequestration did not correlate with the amounts of organic matter inputs (Dexter et al., 1982; Campbell et al., 1991; Johnson and Chamber, 1996; Nicholson et al., 1997), implying that $k_2 C$ exceeded $k_1 A$ or that the rate coefficients were not constant over the duration of these experiments. The k_1 coefficient is a function of C input quality (Franck et al., 1997; Heal et al., 1997; Wang et al., 2004; Johnson et al., 2007a). The k_2 coefficient is affected by tempera-ture, rainfall, soil texture, mineralogy, and residue management, especially tillage (Bayer et al., 2006). Although first-order kinetics can provide preliminary informa-tion, the rates of decomposition and humification slow as more labile materials are decomposed (Wieder and Lang, 1982; Johnson et al., 2007a).

When above-ground biomass is harvested, the quality and quantity of C inputs change because roots and other plant organs may have different chemical composi-tions (Johnson et al., 2007a). This has the potential to shift the rate of decomposi-tion and subsequent humification (k_1). Carbon originating from root biomass and rhizodeposition contributes 1.5 to 3.0 times more C to stable SOM compared to C originating from above-ground biomass (Balesdent and Balabane, 1996; Allmaras et al., 2004; Wilts et al., 2004; Hooker et al., 2005). The higher values correspond to systems with little or no incorporation of shoot material. Unincorporated residues decompose more slowly (Ghidey and Alberts, 1993) and have fewer opportunities to enter the soil. Roots of corn, alfalfa (*Medicago sativa* L.), and switchgrass (*Panicum virgatum* L.) decompose more slowly than corresponding leaves or stems, but this is not the case for soybean or cuphea (*Cuphea viscisissima* Jacq. and *Cuphea lanceolata* W.T. Aiton) (Johnson et al., 2007a). Although roots contribute more C to SOC, they comprise less plant biomass than above-ground biomass for most annual species (Amos and Walters, 2006; Johnson et al., 2006a).

Using empirical data and linear regression of C inputs and SOC, Johnson et al. (2006a) proposed *minimum source C* (MSC) as a term to describe annual C inputs necessary for dC/dt (Equation 1.3) to equal zero, implying no net change in SOC content. For many agronomic crops, grain is harvested and not returned to the soil, and thus is not included in calculating MSC. Since the Johnson et al. (2006a) review, several other studies allowing MSC estimates revealed similar above-ground MSC estimates (Table 1.2). Using above-ground non-grain C inputs, MSC was 2.5 ± 1.7 Mg C ha^{-1} yr^{-1} (n = 28) for different crops and tillage practices at several experimental sites—slightly higher than the mean MSC of 2.2 ± 1.1 Mg C ha^{-1} yr^{-1} (n = 21) cited by Johnson et al. (2006a).

Moldboard plowed systems had higher MSC requirements than those with no tillage; this result was also reported by Bayer et al. (2006). In general, wheat systems have lower MSC than corn-based systems (Kong et al., 2005; Sainju et al., 2006; Kundu et al., 2007). When rhizodeposition is included, MSC values are larger (Clay et al., 2006; Huggins et al., 2007).

Herbaceous perennial species (e.g., switchgrass) have extensive and deep rooting systems (Ma et al., 2000), and thus may exhibit low above-ground MSC relative to annual species so long as sufficient cover is provided to minimize erosion. Several authors reported increases in SOC under perennial grasses. After six years, SOC under tall fescue (*Festuca arundinacea*) was 3 Mg ha^{-1} greater than under corn in Ohio (Lal et al., 1998). After four years, SOC under switchgrass stands in southwestern Quebec increased by 3 Mg ha^{-1} compared to corn (Zan et al., 2001). In a three-year study, SOC increased at 10 Mg C ha^{-1} yr^{-1} (0 to 0.9 m depth) in central North Dakota under switchgrass harvested annually (Frank et al., 2004). The very low initial soil C content of the North Dakota soil was thought to contribute to the very high SOC accrual rate.

The MSC is a useful guideline for determining the amount of allowable biomass harvest for a management system (Johnson et al., 2006a; Johnson et al., 2006b; Wilhelm et al., 2007). Clearly, given the range of MSC values reported, using an average value is unlikely to provide accurate local harvest rates. Improved understanding of SOM dynamics is critical to developing sustainable biomass harvest guidelines. In the short term, use of process or mechanistic models such as CENTURY (Parton, 1996) or CQSTR (Rickman et al., 2002) may be useful to estimate site- and system-specific biomass harvest rates.

TABLE 1.2

Empirical Estimates of Annual Above-Ground Non-Grain Carbon Inputs Required to Maintain Soil Organic Carbon Levels

Location	Crop*	Primary Tillage**	Soil Type§	MSC (Mg C ha⁻¹ yr⁻¹)	Citation
MN	M	CP	SiL	2.6	Allmaras et al., 2004
SD	M	CP	L	3.21	Pikul et al., 2008
WI	M	NT	SiL	2.0	Kucharik et al., 2001
MI	M	MBP	SaL	1.6	Vitosh et al., 1997
WI	M	MBP	SiL	2.3	Vanotti et al., 1997
IN	M	MBP	SiL	>4.0	Barber, 1979
IA	M	MBP	CL	2.4	Larson et al., 1972
MN	M	MBP	CL	3.0	Crookston et al., 1991 Huggins et al., 1998
MN	M	MBP	Cl, SiCL, SiL	3.3	Reicosky et al., 2002 Wilts et al., 2004
MN	M, S	MBP	CL	3.0	Crookston et al., 1991 Huggins et al., 1998
NE	M, S	D	SiL	2.4	Varvel and Wilhelm, 2008
MN	M, S	NT	CL	8.7	Huggins et al., 2007
SD	M, S	NT or ST	CL, L, SiCL	1.6	Clay et al., 2001; 2006
KS	S, Sr	CP	SiL	1.7	Havlin and Kissel, 1997
KS	S, Sr	NT	SiL	1.2	Havlin and Kissel, 1997
WA	W	NR	—	1.2	Horner et al., 1960 Rasmussen et al., 1980
KS	W	NR	—	2.0	Hobbs and Brown, 1965; Rasmussen, 1980
OR	W	MBP	SiL	2.1	Rasmussen, 1980
WA	W	NR	SiL	2.0	Paustian et al., 1997
WA	W	NR	SiL	4.0	Paustian et al., 1997
MT	W	NT	CL	0.82	Sainju et al., 2006
MT	W	V	SaL	0.3	Black, 1973
CA	W, M, T	CT	SiL, SiCL	2.6	Kong et al., 2005
Sweden	W, Ba	HT	SaCL	1.5	Paustian et al., 1992
Mexico	W, M	MBP, NT	C	1.5	Follett et al., 2005
India	W, S	NR	SaL	0.032	Kundu et al., 2007
Brazil	O, M, V, C	CT	SaCL	6.2	Bayer et al., 2006
Brazil	O, M, V, C	NT	SaCL	2.7	Bayer et al., 2006
Average				2.5 ± 1.7	N = 28

* Crops: Ba = barley (*Hordeum vulgare* L.). C = cowpea (*Vigna unguiculta* (L.) Walp.). M = maize (*Zea Mays* L.). O = oat (*Avena strigoas* Schreb.). S = soybean (*Glycine max* (L.) Merr.). Sr = sorghum (*Sorghum bicolor* L.). T = tomato (*Lycopersicon esculentum* Mill.). V = vetch (*Vicia sativa* L.). W = wheat (*Triticum aestivum* L.).

** Primary tillage: CP = chisel plow. CT = conventional tillage, details not provided. D = disk. HT = hand tillage. MBP = moldboard plow. NR = not reported. RT = ridge till. ST = strip tillage. V = V-blade, 9 to12 cm.

§ Soil type: Si = silt. Sa = sandy. L = loam. C = clay.

SOIL BIOLOGY AND NUTRIENT CYCLING

Plant biomass provides a complex matrix of organic materials that interact with SOM. This complexity can influence the diversity of the soil microbial community and related physiological and enzymatic processes, thus affecting nutrient mineralization and N availability (Bending et al., 2002). In general, harvesting crop residues reduces indicators of soil biology activity. Based on a compendium of worldwide studies, harvesting residues reduced the concentrations of microbial biomass C by 25% and microbial biomass N by 29% (Tables 1.3 and 1.4). In some cases, the impact of residue harvest was measured as early as two years after biomass harvest. Of the 25 observations, only three cases indicated that residue removal had no effect or exerted a positive impact on microbial biomass C concentration relative to treatments that retained residue (Table 1.3). Microbial biomass C increased proportionally to the amount of biomass returned when harvest rates were varied (Karlen et al., 1994; Cookson et al., 1998; Debosz et al., 1999; Salinas-Garcia et al., 2001; Limon-Ortega et al., 2006).

Earthworms are macroscopic indicators of a healthy soil and provide beneficial functions related to nutrient cycling, soil structure, hydrology, and root growth. A reduction in earthworm activity caused a decrease in saturated hydraulic conductivity to a depth of 20 cm (Blanco-Canqui et al., 2007). Numerous studies have noted that reducing or eliminating tillage increases earthworm biomass (Nuutinen, 1992) and abundance (Edwards et al., 1990; Nuutinen, 1992; Kladivko, 2001). Elimination of burning crop residue also increased earthworm abundance (Fraser et al., 1996; Wuest et al., 2005). Therefore, it was expected that retaining non-grain biomass would reveal greater earthworm populations than areas from which the biomass was removed (Table 1.5). Similar to microbial biomass C, earthworm abundance increased with the amount of biomass returned when harvest rates varied such that at least 25% of corn stover decreased midden numbers.

Crop residue management may influence plant diseases. Retention of residues can result in net changes in soil microbiota by retaining inoculum or creating an environment more conducive to pathogens (Cook et al., 1978). For example, corn stover and small grain straws are the principal inoculum sources of *Fusarium* spp. that cause head blight in wheat, especially in no-till systems (Maiorano et al., 2008). Population counts of *Fusarium* were highest in continuous corn with residue retention and lowest under continuous wheat or corn–wheat rotation, also with residue retained (Govaerts et al., 2008). The same study indicated that residue retention increased populations of disease-suppressing microorganisms including fluorescent *Pseudomonas* that provide biological control of *Fusarium* and other fungal pathogens.

Greater incidence of root rot in corn was associated with stover retention relative to stover harvest without tillage; however, root rot did not reduce yield (Govaerts et al., 2007a). Govaerts et al. (2008) proposed that no tillage and residue retention have potential for biological control by promoting plant growth and suppressing disease, but eliminating tillage alone did not improve soil health. Disease response to repeated burning of wheat stubble was variable, depending on precipitation and N management (Smiley et al., 1996). Crown rot incidents were positively correlated with SOC, while root rot incidents were negatively correlated to microbial biomass

TABLE 1.3

Microbial Biomass C Concentration (µg g⁻¹) or Content (kg ha⁻¹) Response to Non-Grain Biomass Harvest Treatment

Location	Soil*	Crop**	Tillage§	Years	Depth (cm)	Biomass Treatment		Comment	Citation¶
						Removed	Retained		
								µg Microbial biomass C g⁻¹	
Lancaster, WI	SiL	M, M	NT	10	0 to 8	330	696	Normal retained	b
Lancaster, WI	SiL	M, M	NT	10	0 to 8	330	1060	Double biomass	b
Ste-Anne-de-Bellevue, Que, Canada	Sa-LSa	M, M	MBP, RT, NT	9	0 to 10	124	200	Average across tillage	h
Ste-Anne-de-Bellevue, Que, Canada	S-LSa	M, M	MBP, RT, NT	9	10 to 20	128	158	Average across tillage	h
El Batán, Mexico	L	M, M	NT	14	0 to 15	122	453		j
El Batán, Mexico	L	M, M	CT	14	0 to 15	291	322		j
El Batán, Mexico	L	W, W	NT	14	0 to 15	329	374		j
El Batán, Mexico	L	W, W	CT	14	0 to 15	288	544		j
Central Mexico	SaL	M, M	NT	5	0 to 15	303	654	7 Mg ha⁻¹ retained	g
Central Mexico	SaL	M, M	NT	5	0 to 15	303	426	5 Mg ha⁻¹ retained	g
Central Mexico	SaL	M, M	NT	5	0 to 15	303	345	3 Mg ha⁻¹ retained	g
Central Mexico	SaL	M, M	NT	5	0 to 15	303	495	3 Mg ha⁻¹ retained + vicia cover	g
Central Mexico	SaL	M, M	NT	5	0 to 15	303	488	3 Mg ha⁻¹ retained + *P. vulgaris* cover	g
Central Mexico	SaL	M, M	CT	5	0 to 15	NR#	264	All biomass retained	g
Yaqui Valley, Mexico	SaL	W, M	CT	3	0 to 7	NR	425	All retained	f
Yaqui Valley, Mexico	SaL	W, M	CT	3	7 to 15	NR	292	All retained	f

TABLE 1.3 (CONTINUED)

Microbial Biomass C Concentration (µg g⁻¹) or Content (kg ha⁻¹) Response to Non-Grain Biomass Harvest Treatment

Location	Soil*	Crop**	Tillage§	Years	Depth (cm)	Biomass Treatment		Comment	Citation[1]
						Removed	Retained		
						µg Microbial biomass C g⁻¹			
Yaqui Valley, Mexico	SaL	W, M	NT	3	0 to 7	321	350	Only stover harvested, wheat straw retained	f
Yaqui Valley, Mexico	SaL	W, M	NT	3	7 to 15	218	184	Only stover harvested, wheat straw retained	f
Yaqui Valley, Mexico	SaL	W, M	NT	3	0 to 7	321	348	Stover and straw harvested	f
Yaqui Valley, Mexico	SaL	W, M	NT	3	7 to 15	218	210	Stover and straw harvested	f
Yaqui Valley, Mexico	SaL	W, M	CT	10	0 to 7	NR	461	All retained	f
Yaqui Valley, Mexico	SaL	W, M	NT	10	0 to 7	584	642	Only wheat straw retained	f
Yaqui Valley, Mexico	SaL	W, M	NT	10	0 to 7	584	533	Both retained	f
Yaqui Valley, Mexico	SaL	W, M	CT	12	0 to 7	NR	596	All retained	f
Yaqui Valley, Mexico	SaL	W, M	CT	12	0 to 7	617	681	Only stover harvested, wheat straw retained	f
Yaqui Valley, Mexico	SaL	W, M	NT	12	0 to 7	617	687	Stover and straw harvested	f
Nairobi, Kenya	Humic nitisol	C, B	CU	18	0 to 15	54	72		c
Varanasi, India	SaL	R, Ba	D*2, CU	2	0 to 10	235	347	Cultivated to 20 cm	d
Varanasi, India	SaL	R, Ba	D, CU	2	0 to 10	271	427	Cultivated to 10 cm	d
Varanasi, India	SaL	R, Ba	NT	2	0 to 10	283	320	Averaged across crop periods	d
						kg Microbial Biomass C ha⁻¹			
Apatzingan, Mexico	C	M, M	NT	6	0 to 5	830	970	100% stover retained	e

Apatzingan, Mexico	C	M, M	NT	6	5 to 10	705	760	100% stover retained	e
Apatzingan, Mexico	C	M, M	NT	6	10 to 20	640	705	100% stover retained	e
Apatzingan, Mexico	C	M, M	NT	6	0 to 5	830	830	66% stover retained	e
Apatzingan, Mexico	C	M, M	NT	6	5 to 10	705	760	66% stover retained	e
Apatzingan, Mexico	C	M, M	NT	6	10 to 20	640	708	66% stover retained	e
Apatzingan, Mexico	C	M, M	NT	6	0 to 5	830	805	33% stover retained	e
Apatzingan, Mexico	C	M, M	NT	6	5 to 10	705	750	33% stover retained	e
Apatzingan, Mexico	C	M, M	NT	6	10 to 20	640	680	33% stover retained	e
Apatzingan, Mexico	C	M, M	MBP	6	0 to 5	NR	710	100% stover retained	e
Apatzingan, Mexico	C	M, M	MBP	6	5 to 10	NR	815	100% stover retained	e
Apatzingan, Mexico	C	M, M	MBP	6	10 to 20	NR	640	100% stover retained	e
Apatzingan, Mexico	C	M, M	D	6	0 to 5	NR	810	100% stover retained	e
Apatzingan, Mexico	C	M, M	D	6	5 to 10	NR	760	100% stover retained	e
Apatzingan, Mexico	C	M, M	D	6	10 to 20	NR	645	100% stover retained	e
Casas Blancas, Mexico	SiC	M, M	NT	6	0 to 5	350	710	100% stover retained	e
Casas Blancas, Mexico	SiC	M, M	NT	6	5 to 10	310	500	100% stover retained	e
Casas Blancas, Mexico	SiC	M, M	NT	6	10 to 20	300	360	100% stover retained	e
Casas Blancas, Mexico	SiC	M, M	NT	6	0 to 5	350	650	66% stover retained	e
Casas Blancas, Mexico	SiC	M, M	NT	6	5 to 10	310	490	66% stover retained	e
Casas Blancas, Mexico	SiC	M, M	NT	6	10 to 20	300	300	66% stover retained	e
Casas Blancas, Mexico	SiC	M, M	NT	6	0 to 5	350	500	33% stover retained	e
Casas Blancas, Mexico	SiC	M, M	NT	6	5 to 10	310	480	33% stover retained	e
Casas Blancas, Mexico	SiC	M, M	NT	6	10 to 20	300	280	33% stover retained	e
Casas Blancas, Mexico	SiC	M, M	MBP	6	0 to 5	NR	300	100% stover retained	e
Casas Blancas, Mexico	SiC	M, M	MBP	6	5 to 10	NR	280	100% stover retained	e
Casas Blancas, Mexico	SiC	M, M	MBP	6	10 to 20	NR	240	100% stover retained	e
Casas Blancas, Mexico	SiC	M, M	D	6	0 to 5	NR	475	100% stover retained	e
Casas Blancas, Mexico	SiC	M, M	D	6	5 to 10	NR	440	100% stover retained	e

TABLE 1.3 (CONTINUED)
Microbial Biomass C Concentration (µg g⁻¹) or Content (kg ha⁻¹) Response to Non-Grain Biomass Harvest Treatment

Location	Soil*	Crop**	Tillage[s]	Years	Depth (cm)	Biomass Treatment Removed	Retained	Comment	Citation[1]
						kg Microbial Biomass C ha⁻¹			
Casas Blancas, Mexico	SiC	M, M	D	6	10 to 20	NR	350	100% stover retained	e
Biloela, Australia	SaC	Sr	D, CU	6	12	273	315		a
Biloela, Australia	SaC	Sr	NT	6	10	313	347		a

* Soils: Si = silt, silty. C = clay. L = loam. Sa = sand, sandy. V = vertisol. H = haplustol.

** Crops: Ba = barley (*Hordeum vulgare* L.). B = bean (*Phaseolus* L.). G = grass. M = maize (*Zea Mays* L.). R = rice (*Oryza sativa* L.). Sr = sorghum (*Sorghum bicolor* L.). W = wheat (*Triticum aestivum* L.).

[s] Primary tillage: CT = conventional tillage, implement not designated. CU = cultivated. D = disk. MBP = moldboard plow. NT = no tillage. RT = ridge tillage.

[1] Citations: a = Saffigna et al., 1989; b = Karlen et al., 1994; c = Kapkiyai et al., 1999; d = Kushwaha et al., 2000; e = Salinas-Garcia et al., 2001; f = Limon-Ortega et al., 2002; g = Roldan et al., 2003; h = Spedding et al., 2004; i = Limon-Ortega et al., 2006; j = Govaerts et al., 2007b.

NR = not reported or treatment not measured.

TABLE 1.4

Microbial Biomass N Concentration (µg g⁻¹) or Content (kg ha⁻¹) Response to Non-Grain Biomass Harvest

Location	Soil*	Crop**	Tillage§	Years	Depth (cm)	Biomass Treatment		Comment	Citation¶
						Removed	Retained		
						μg Microbial Biomass N g⁻¹			
Canterbury, New Zealand	NR	W-W-Ba-Ba	CU	4	0 to 5	38	47		b
Canterbury, New Zealand	NR	W-W-Ba-Ba	CU	4	5 to 10	34	49		b
Canterbury, New Zealand	NR	W-W-Ba-Ba	CU	4	10 to 25	18	38		b
Ste-Anne-de-Bellevue, Que, Canada	Sa-LSa	M-M	MBP, RT, NT	9	0 to 10	13	26		f
Ste-Anne-de-Bellevue, Que, Canada	Sa-LSa	M-M	MBP, RT, NT	9	10 to 20	19	19		f
El Batán, Mexico	L	M-M	NT	14	0 to 15	17	39		g
El Batán, Mexico	L	M-M	CT	14	0 to 15	12	24		g
El Batán, Mexico	L	W-W	NT	14	0 to 15	21	29		g
El Batán, Mexico	L	W-W	CT	14	0 to 15	25	27		g
Nairobi, Kenya	Humic nitisol	M-B	CU	18	0 to 15	9.4	11.2		c
Varanasi, India	SaL	R-Ba	D*2, CU	2	0 to 10	24	38	Cultivated to 20 cm	d
Varanasi, India	SaL	R-Ba	D, CU	2	0 to 10	28	49	Cultivated to 10 cm	d
Varanasi, India	SaL	R-Ba	NT	2	0 to 10	27	31		d
Biloela, Australia	SaC	Sr	D, CU	5	12	39	48		a
Biloela, Australia	SaC	Sr	NT	5	10	37	46		a
						kg Microbial Biomass N ha⁻¹			
Apatzingan, Mexico	C	M-M	NT	6	0 to 5	33	53	100% stover retained	e
Apatzingan, Mexico	C	M-M	NT	6	5 to 10	34	43	100% stover retained	e

TABLE 1.4 (CONTINUED)
Microbial Biomass N Concentration (µg g⁻¹) or Content (kg ha⁻¹) Response to Non-Grain Biomass Harvest

Location	Soil*	Crop**	Tillage§	Years	Depth (cm)	Biomass Treatment		Comment	Citation¶
						Removed	Retained		
						kg Microbial Biomass N ha⁻¹			
Apatzingan, Mexico	C	M-M	NT	6	10 to 20	28	34	100% stover retained	e
Apatzingan, Mexico	C	M-M	NT	6	0 to 5	33	44	66% stover retained	e
Apatzingan, Mexico	C	M-M	NT	6	5 to 10	34	40	66% stover retained	e
Apatzingan, Mexico	C	M-M	NT	6	10 to 20	28	33	66% stover retained	e
Apatzingan, Mexico	C	M-M	NT	6	0 to 5	33	39	33% stover retained	e
Apatzingan, Mexico	C	M-M	NT	6	5 to 10	34	37	33% stover retained	e
Apatzingan, Mexico	C	M-M	NT	6	10 to 20	28	32	33% stover retained	e
Apatzingan, Mexico	C	M-M	MBP	6	0 to 5	NR#	24	100% stover retained	e
Apatzingan, Mexico	C	M-M	MBP	6	5 to 10	NR	25	100% stover retained	e
Apatzingan, Mexico	C	M-M	MBP	6	10 to 20	NR	24	100% stover retained	e
Apatzingan, Mexico	C	M-M	D	6	0 to 5	NR	46	100% stover retained	e
Apatzingan, Mexico	C	M-M	D	6	5 to 10	NR	35	100% stover retained	e
Apatzingan, Mexico	C	M-M	D	6	10 to 20	NR	29	100% stover retained	e
Casas Blancas, Mexico	SiC	M-M	NT	6	0 to 5	41	72	100% stover retained	e
Casas Blancas, Mexico	SiC	M-M	NT	6	5 to 10	38	42	100% stover retained	e
Casas Blancas, Mexico	SiC	M-M	NT	6	0 to 20	34	35	100% stover retained	e
Casas Blancas, Mexico	SiC	M-M	NT	6	0 to 5	41	60	66% stover retained	e
Casas Blancas, Mexico	SiC	M-M	NT	6	5 to 10	38	41	66% stover retained	e
Casas Blancas, Mexico	SiC	M-M	NT	6	10 to 20	34	35	66% stover retained	e
Casas Blancas, Mexico	SiC	M-M	NT	6	0 to 5	41	51	33% stover retained	e
Casas Blancas, Mexico	SiC	M-M	NT	6	5 to 10	38	39	33% stover retained	e

Casas Blancas, Mexico	SiC	M-M	NT	6	10 to 20	34	35	33% stover retained	e
Casas Blancas, Mexico	SiC	M-M	MBP	6	0 to 5	NR	40	100% stover retained	e
Casas Blancas, Mexico	SiC	M-M	MBP	6	5 to 10	NR	32	100% stover retained	e
Casas Blancas, Mexico	SiC	M-M	MBP	6	10 to 20	NR	33	100% stover retained	e
Casas Blancas, Mexico	SiC	M-M	D	6	0 to 5	NR	50	100% stover retained	e
Casas Blancas, Mexico	SiC	M-M	D	6	5 to 10	NR	33	100% stover retained	e
Casas Blancas, Mexico	SiC	M-M	D	6	10 to 20	NR	31	100% stover retained	e

* Soils: Si = silt, silty. C = clay. L = loam. Sa = sand, sandy. V = vertisol. H = haplustol.

** Crops: Ba = barley (*Hordeum vulgare* L.). B = bean (*Phaseolus*). G = grass. M = maize (*Zea mays* L.). R = rice (*Oryza sativa* L.). Sr = sorghum (*Sorghum bicolor* L.). W = wheat (*Triticum aestivum* L.).

§ Tillage: CT = conventional tillage, implement not designated. CU = cultivated. D = disk. MBP = moldboard plow. NT = no tillage. RT = ridge tillage.

¶ Citations: a = Saffigna et al., 1989; b = Fraser and Piercy, 1998; c = Kapkiyai et al., 1999; d = Kushwaha et al., 2000; e = Salinas-Garcia et al., 2001; f = Spedding et al., 2004; g = Govaerts et al., 2007b.

NR = not reported or treatment not measured.

TABLE 1.5
Earthworm Response to Non-Grain Biomass Harvest Treatment

Location	Soil*	Crop**	Tillage§	Years	Depth (cm)	Earthworm Abundance Biomass Removed	Biomass Retained	Comment	Citation¶
							(Number m⁻²)		
Jokioinen, Finland	SiCL	SW	MBP, CU	8/9	NA	14	17	Average tillage and years	a
Mouhijarvi, Finland	SaCL	SW	MBP, CU	8/9	NA	18	19	Average across tillage	a
Palkane, Finland	L	SW	MBP, CU	8/9	NA	16	20	Average across tillage	a
Lancaster, WI	SiL	M-M	NT	10	NA	53	78		b
Shanxi, China	SiL	WiW	MBP	6	30	0	NR		e
Shanxi, China	SiL	WiW	NT	6	30	NR	5		e
Shanxi, China	SiL	WiW	MBP	14	30	0	NR		e
Shanxi, China	SiL	WiW	NT	14	30	NR	19		e
Canterbury, New Zealand	NR	W-W-Ba-Ba	M	4	25	NR	400		c
Canterbury, New Zealand	NR	W-W-Ba-Ba	CU	4	25	125	NR		c
							Number middens m⁻²		
Coshocton, OH	SiL	M	NT	1	Surface	3	19	10 Mg ha⁻¹ stover retained	d
Coshocton, OH	SiL	M	NT	1	Surface	3	10	5 Mg ha⁻¹ stover retained	d
Coshocton, OH	SiL	M	NT	1	Surface	3	5	3.75 Mg ha⁻¹ stover retained	d
Coshocton, OH	SiL	M	NT	1	Surface	3	5	2.5 Mg ha⁻¹ stover retained	d
Coshocton, OH	SiL	M	NT	1	Surface	3	5	1.25 Mg ha⁻¹ stover retained	d
South Charleston, OH	SiL	M	NT	1	Surface	1	17	10 Mg ha⁻¹ stover retained	d
South Charleston, OH	SiL	M	NT	1	Surface	1	17	5 Mg ha⁻¹ stover retained	d
South Charleston, OH	SiL	M	NT	1	Surface	1	12	3.75 Mg ha⁻¹ stover retained	d

South Charleston, OH	SiL	M	NT	1	Surface	1	4	2.5 Mg ha^{-1} stover retained	d
South Charleston, OH	SiL	M	NT	1	Surface	1	3	1.25 Mg ha^{-1} stover retained	d
Hoytville, OH	CL	M	NT	1	Surface	2	24	10 Mg ha^{-1} stover retained	d
Hoytville, OH	CL	M	NT	1	Surface	2	18	5 Mg ha^{-1} stover retained	d
Hoytville, OH	CL	M	NT	1	Surface	2	15	3.75 Mg ha^{-1} stover retained	d
Hoytville, OH	CL	M	NT	1	Surface	2	13	2.5 Mg ha^{-1} stover retained	d
Hoytville, OH	CL	M	NT	1	Surface	2	10	1.25 Mg ha^{-1} stover retained	d

* Soils: Si = silt, silty. C = clay. L = loam. Sa = sand, sandy. V = vertisol. H = haplustol.

** Crops: Ba = barley (*Hordeum vulgare* L.). B = bean (*Phaseolus*). G = grass. M = maize (*Zea mays* L.). R = rice (*Oryza sativa* L.). S = sorghum (*Sorghum bicolor* L.). W = wheat (*Triticum aestivum* L.).

§ Tillage: CT = conventional tillage, implement not designated. CU = cultivated. D = disk. MBP = moldboard plow. NT = no tillage. RT = ridge tillage.

¶ Citations: a = Nuutinen, 1992; b = Karlen et al., 1994; c = Fraser and Piercy, 1998; d = Blanco-Canqui et al., 2007; e = Li et al., 2007.

NA = not applicable, soil treated to draw worms to soil surface.

NR = not reported or treatment not measured.

(Smiley et al., 1996). Effects of residue harvest on microbial species are difficult to predict; negative, neutral, and positive responses have been reported along with interactions of tillage, climate, and nutrient management.

Harvesting crop non-grain biomass affects nutrient cycling by removing plant macronutrients (N, P, K, Ca, and Mg) (Mubarak et al., 2002) and micronutrients (Fageria, 2004). The concentration of nutrient in non-grain biomass averaged 9.0 ± 5.1 g N kg^{-1}, 1.1 ± 0.52 g P kg^{-1}, and 5.0 ± 1.3 g K kg^{-1} based on results from several common annual crops and likely perennial biomass crops (Table 1.6). The amounts of nutrients removed vary among plant species, organs harvested (cob versus entire stover), physiological stage, and amount of biomass harvested (Lindstrom, 1986; Burgess et al., 2002; Mubarak et al., 2002; Fageria, 2004; Johnson et al., 2007a).

The amount of nutrient removed can be calculated from the concentration and biomass harvest rates. Bransby et al. (1998) indicated that harvest of above-ground switchgrass biomass has the potential to remove 126 to 281 kg N ha^{-1} or more under fertilized conditions and 38 kg N ha^{-1} under unfertilized conditions. From a nutrient management view, harvesting biomass after senescence removes the least amount of mineral nutrient. In Washington state, the amount of N removed by harvesting switchgrass varied by cultivar and harvest date more than by the amount of N fertilizer applied; early harvesting prior to N translocation below ground removed more N than harvesting in October (personal communication, Hal Collins, USDA ARS, Prosser, WA). Continued removal of nutrient without replacement by applying fertilizer, manure, or compost depletes soil fertility, in turn reducing soil productivity.

Harvesting non-crop biomass affects soil microbial processes that impact N availability. For example, the activity of N-acetyl-b-D-glucosaminidase was reduced by harvesting corn stover for 10 years in a continuous corn system, suggesting a reduction in N mineralization (Ekenler and Tabatabai, 2003). In Kenya, corn stover harvest for 18 years in a corn–bean (*Phaseolus vulgaris* L.) rotation reduced N stocks (0 to 15 cm depth) that corresponded to declines in total N, particulate matter N, mineral N, microbial biomass N, and potentially mineralizable N (Kapkiyai et al., 1999). Corn and soybeans took up more N where stover was retained than where stover was removed, possibly because stover maintained a soil environment more conducive to biological activity that increased N availability (Power et al., 1986). In India, more N was available (0 to 30 cm depth) with residue retained and incorporated compared to residue removal in wheat–groundnut (*Archis hypogea* L.) rotation (Bhatnagar et al., 1983). These studies indicate increased plant-available N with stover retention and suggest that harvesting non-grain biomass may impact soil fertility adversely.

SOIL AGGREGATION

In agricultural systems, maintenance of SOM has long been recognized as a strategy to improve soil structure and reduce soil degradation. Soil structure is an important property that mediates many physical and biological processes and controls SOM and residue decomposition (Van Veen and Kuikman, 1990). Soil aggregates are the basic units of soil structure and consist of primary particles and binding agents (Figure 1.2; Edwards and Bremner, 1967; Tisdall and Oades, 1982; Tisdall, 1996; Jastrow and Miller, 1997). Water stability of soil aggregates depends on organic

TABLE 1.6

Plant Concentration ± Standard Deviations Based on Literature Reports of N, P, and K in Non-Grain Above-Ground Portions of Potential Non-Grain Biomass Feedstocks

Crop	N (g/kg)	P (g/kg)	K (g/kg)	Citations*
Annuals				
Barley (*Hordeum vulgare* L.)	6.5 ± 1.1	1.1 ± NA[‡]	12.5 ± NA	a
N	6	1	1	
Maize (*Zea mays* L.)	7.5 ± 3.0	1.3 ± 0.5	11.8 ± 6.2	b
N	16	6	5	
Millet (*Panicum miliaceaum* L.)	8.9 ± 5.6	0.85 ± 0.21	12.8 ± NA	c
N	3	2	1	
Rice (*Oryza sativa* L.)	9.3 ± 7.2	0.71 ± 0.24	19.3 ± 14.1	d
N	9	4	3	
Sorghum (*Sorghum bicolor* L.)	12.0 ± 9.4	0.5 ± NA	NR[§]	e
N	4	1		
Soybean (*Glycine max* (L.)	17.6 ± 12.1	1.85 ± 0.5	12.8 ± 3.3	f
N	11	2	2	
Wheat (*Triticum aestivum* L.).	6.8 ± 2.2	1.1 ± 0.6	7.8 ± 2.7	g
N	10	2	2	
Perennials				
Miscanthus (*Miscanthus* × *giganteus*)	8.0 ± 5.3	0.25 ± 0.21	3.45 ± 0.21	h
N	6	2	2	
Switchgrass (*Panicum virgatum* L.)	6.8 ± 4.2	0.62 ± 0.23	3.2 ± 3.3	i
N	17	7	6	
Other grass[¶]	9.0 ± 5.1	1.08 ± 0.52	5.0 ± 1.3	j
N	5	5	5	
Annuals	9.8 ± 7.5	1.1 ± 0.5	13.1 ± 7.6	
N	60	18	14	
Perennials	7.5 ± 4.5	0.73 ± 0.5	3.9 ± 2.4	
N	28	14	13	

*Citations:

a = Christensen, 1986; Lindstrom, 1986; Andren and Paustian, 1987; Cookson et al., 1998; Mitchell et al., 2001; Velthof et al., 2002; Halvorson and Reule, 2007.

b = Lindstrom, 1986; Breakwell and Turco, 1989; Tian et al., 1992; Burgess et al., 2002; Manlay et al., 2002; Velthof et al., 2002; Fageria, 2004; Al-Kaisi et al., 2005; Hoskinson et al., 2007; Yu et al., 2008; Halvorson and Johnson, in press.

c = Manlay et al., 2002; Fatondji et al., 2006; Sarr et al., 2008.

d = Tian et al., 1992; Ying et al., 1998; Manlay et al., 2002; Abiven et al., 2005; Tirol-Padre et al., 2005; Linquist et al., 2007; Kaewpradit et al., 2008.

e = Saffigna et al., 1989; Franzluebbers et al., 1995; Abiven et al., 2005; Monti et al., 2008.

f = Lindstrom, 1986; Franzluebbers et al., 1995; Fageria, 2004; Abiven et al., 2005; Al-Kaisi et al., 2005; Rao et al., 2005; Johnson et al., 2007a.

TABLE 1.6 (CONTINUED)

Plant Concentration ± Standard Deviations Based on Literature Reports of N, P, and K in Non-Grain Above-Ground Portions of Potential Non-Grain Biomass Feedstocks

*Citations (continued):

g = Jawson and Elliott, 1986; Lindstrom, 1986; Franzluebbers et al., 1995; Cookson et al., 1998; Mitchell et al., 2001; Borie et al., 2002; Velthof et al., 2002; Abiven et al., 2005; Tirol-Padre et al., 2005.

h = Clifton-Brown and Lewandowski, 2002; Monti et al., 2008.

i = Bransby et al., 1998; Madakadze et al., 1999; Reynolds et al., 2000; Duffy and Nanhou, 2001; Lemus et al., 2002; Vogel et al., 2002; Cassida et al., 2005; Adler et al., 2006; Lemus et al., 2008; Monti et al., 2008.

j = Katterer et al., 1998; Monti et al., 2008.

NA = not appropriate.

NR = not reported.

¶ Other grasses include cardoon (*Cynara Cardunculus* L.), giant reed (*Arundo donax* L.), and reed canary grass (*Phalaris arundinacea* L.).

---

materials such as polysaccharides, roots, fungal hyphae, and aromatic compounds (Tisdall and Oades, 1982).

SOM is considered a major bonding agent responsible for the formation and stabilization of soil aggregates (Tisdall and Oades, 1982; Dormaar, 1983; Chaney and Swift, 1984; Miller and Jastrow, 1990; Haynes et al., 1991; Degens, 1997; Angers, 1998). In addition, improvement of soil aggregate stability results from the microbial utilization of carbohydrates and from plant phenolics released during decomposition

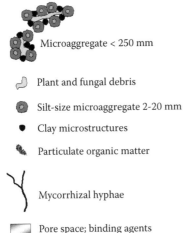

- Microaggregate < 250 mm
- Plant and fungal debris
- Silt-size microaggregate 2-20 mm
- Clay microstructures
- Particulate organic matter
- Mycorrhizal hyphae
- Pore space; binding agents

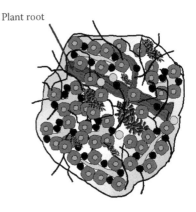

Plant root

Macroaggregate > 250 mm

FIGURE 1.2   Soil macroaggregate formation. (*Source:* Jastrow and Miller, 1997. In Lal, R. et al., Eds., *Soil Processes and the Carbon Cycle.* CRC Press, Boca Raton, FL, pp. 207–223.)

of structural components such as lignin (Martens, 2000). Plant root and fungal hyphae form a network in soil that entangles microaggregates to form macroaggregates that are then stabilized by extracellular polysaccharides that confer increased resilience to aggregates in the presence of water (Elliott and Coleman, 1988; Tisdall, 1994). The length of fungal hyphae can be reduced by harvesting residue (Cookson et al., 1998), which may contribute to a reduction in aggregate stability.

The addition of fresh organic residue induces the formation and stabilization of macroaggregates by the addition of a C source for microbial activity (Golchin et al., 1994b; Jastrow, 1996; Six et al., 1999; Mikha and Rice, 2004; Johnson et al., 2007c). In a conceptual model proposed by Golchin et al. (1994a), plant residues are colonized by microorganisms as they enter the soil. Plant fragments also can be encrusted by mineral particles that become the centers of water-stable aggregates (Figure 1.2). Since these plant fragments are rich in readily decomposable carbohydrates, microbial metabolites permeate the coatings of mineral particles and stabilize the aggregates (Golchin et al., 1994a). In addition, soil conditions can cause increased solubility of some polyvalent cations such as Fe and Mn, thereby contributing to the formation of soil microaggregates and the stabilization of SOM (Figure 1.3) through formation of cation bridges (Elliott and Coleman, 1988). Thus, the addition of organic residues high in available C can promote the stabilization of soil aggregates; conversely, insufficient C inputs can lead to losses of stable aggregates.

Different management practices affect formation and stabilization of soil aggregates through their effects on SOM level and soil biota (Tisdall and Oades, 1982; O'Halloran et al., 1986; Beare and Bruce, 1993; Edwards et al., 1993; Frey et al., 1999; Six et al., 2000a). Cultivation affects soil structure due to the destruction of soil aggregates and the loss of SOM (Low, 1972; Van Veen and Paul, 1981; Tisdall

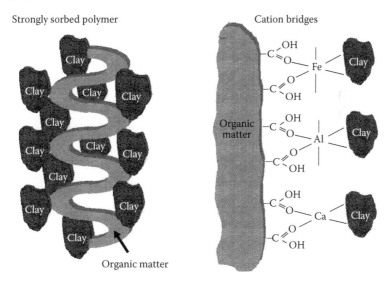

FIGURE 1.3 Soil microaggregate formation (<250 μm) and SOM stabilization. Note cation bridges that connect SOM and clay particles. (*Source:* Tisdall, J.M. and J.M. Oades. 1982. *J. Soil Sci.* 33:141–163.)

and Oades, 1982; Elliott, 1986; Angers et al., 1992; Six et al., 1998; Six et al., 1999). Losses of SOM from cultivation of grassland have been attributed at least partly to the mineralization of organic materials that bind microaggregates into macroaggregates (Elliott, 1986; Gupta and Germida, 1988).

McVay et al. (2006) observed that aggregate stability was greatest in treatments with the highest SOC in several long-term studies in Kansas. De Gryze et al. (2005) observed that aggregate formation increased linearly with increasing residue amounts at a rate of 12.0 ± 1.24 g aggregate g^{-1} residue added. They also observed that macroaggregates (> 2000 µm) increased from 3% to 40% as the amount of residue added was increased from 0 g to 3 g per 100 g soil. Water-stable aggregation index was significantly greater in a tillage-plus-straw-retained treatment (0.97) compared to tillage without straw treatment (0.68) (Singh et al., 1994). In another study, addition of a high-lignin organic material and corn stover increased water-stable aggregates (Johnson et al., 2007c). These results are indicative of the beneficial effects of organic matter addition on the aggregation process.

Soil aggregation affects soil water and aeration, which are important factors in crop production. The size, shape, and stability of soil aggregates impact pore size distribution (Lynch and Bragg, 1985). Soil structural stability depends on the ability of aggregates to remain intact when subjected to stress such as rapid wetting (Tisdall, 1996). Lynch and Bragg (1985) reported that unstable aggregates slake when wetted. Slaking occurs when aggregates are too unstable to withstand pressures resulting from entrapped air inside air-dried aggregates during rapid rewetting (Elliott, 1986; Gäth and Frede, 1995; Six et al., 2000b). Resistance to slaking is associated with large pieces of organic debris from plant roots, surface litter, and fungal hyphae (Oades, 1984). When air-dried soils are slowly rewetted, changes in aggregates are minimal (Six et al., 2000b). Under field conditions, aggregates near the surface are subjected to more slaking compared to aggregates below the surface layer that are protected from air drying and rapid wetting (Lynch and Bragg, 1985).

Soil aggregation is important for increasing water infiltration. Residue cover protects the soil surface from direct raindrop impact and minimizes aggregate slaking from fast rewetting, thus maintaining soil aggregates and reducing surface crusting compared with bare soil. Unstable aggregates at the surface can lead to the formation of crusts that inhibit water infiltration and air movement into the soil (Tisdall and Oades, 1982; Lynch and Bragg, 1985). Within 24 hours of the formation of surface crusts, the O_2 diffusion rate is reduced by 50% (Rathore et al., 1982). Not tilling and retaining crop stubble increased infiltration rate 3.7-fold compared with conventional tillage (three cultivation passes) and burnt stubble in a 24-year study (Zhang et al., 2007). Water-stable macroaggregates were positively correlated to hydraulic conductivity and negatively correlated to bulk density under dryland crop production in eastern Colorado (Benjamin et al., 2008). Govaerts et al. (2007a) reported that retaining wheat and maize residue improved water infiltration dramatically in both no-till and conventionally tilled plots. Harvest of non-grain biomass has the potential to increase water runoff and soil erosion by impairing soil structure through decreased aggregate stability and macroporosity of the soil surface.

SOIL EROSION

Soil erosion is a two-step process in which soil particles are detached in response to an energy input and then transported. Eroding soil can be moved by wind, water, ice, or gravity. These erosion processes redistribute soil within a landscape and can remove soil under some conditions. Soil loss by erosion removes topsoil from a soil profile. Areas subject to soil loss by erosion are typically less productive because of lower water-holding capacity and decreased fertility. Blowing soil can damage crop plants and soil deposition can bury small plants. From 1994 through 1996, soil erosion caused an annual productivity loss of at least $40 km^{-2} throughout most of the United States Corn Belt; large areas experienced annual losses in excess of $380 km^{-2} (Magleby, 2003). In addition to decreasing productivity, erosion causes off-site impairment of surface water and air quality, property damage, and detrimental effects on human and animal health. Damage caused by soil erosion in the United States has been estimated at $2 billion to $8 billion annually (Magleby, 2003).

The agricultural community recognizes that returning crop residues to the soil is essential to avoid large declines in SOM that adversely impact soil fertility, soil strength, aggregation, and other properties and thus negatively influence crop production. While crop residues affect determinants of soil erodibility (e.g., water-stable aggregates), this section will focus on the physical role of crop residues in reducing soil lost by erosion.

Years of water erosion research consistently indicate for a given soil type, loss rates by water erosion increase with increasing slope gradient, slope length, and rainfall intensity. Soil cover and production practices also influence soil loss via erosion. Crop residue decreases the detachment and transport of soil by water by intercepting raindrops before they impact the soil, by slowing the flow of water over the soil surface, increasing the depth of water on the surface, and by providing small areas of ponded water where sediment can be deposited (Cogo et al., 1982). Soil cover decreases soil loss by water erosion, but the relationship is not linear (Figure 1.4; Lindstrom, 1986; Erenstein, 2002; Merrill et al., 2006). Typically, the ratio of soil loss with a groundcover relative to loss incurred without groundcover decreases exponentially with increasing cover (Figure 1.4). For example, on the same soil type, the amount of soil lost due to water erosion was similar with 100% groundcover and with 60% corn stover or wheat straw cover (Cogo et al., 1982).

Soil loss rates through wind erosion are affected by surface roughness, field dimensions, and wind characteristics. Similar to water erosion, soil cover can decrease soil loss by wind erosion. The effect of soil cover on loss from wind erosion is a complex function of crop type, residue orientation, and other factors. Generally, standing stubble is more effective at reducing wind erosion than flattened residue, and stubble oriented in rows perpendicular to the wind direction is more effective than stubble in rows parallel to the wind direction (Skidmore, 1988). Bilbro and Fryrear (1994) summarized the relationship of soil cover to soil loss by wind erosion. The ratio of the amount of soil lost from protected soil to that lost from flat, bare soil decreased exponentially with increasing groundcover (Soil loss ratio = $e^{(-0.04380*\% \text{ soil cover})}$). This relationship indicates that when soil is at least 50% covered with residue, loss by wind erosion is expected to be 10% or less of losses from flat, bare soil. Wind tunnel

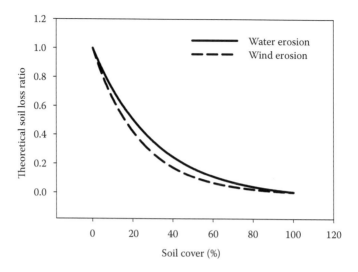

FIGURE 1.4   Soil loss ratios predicted by the revised universal soil loss equation (RUSLE) and revised wind erosion equation (RWEQ) with varying amounts of soil cover. Soil loss ratios include specified amount of residue cover for bare soil under high erodibility conditions. (*Source:* Merrill, S.D. et al. 2006. *J. Soil Water Conserv.* 61:7–13.)

and field studies suggest even low fractions of residue cover can drastically decrease wind erosion (Bilbro and Fryrear, 1994).

For both wind and water erosion, the relationship between soil cover and loss is a non-linear function where >50% groundcover can virtually eliminate erosion (Figure 1.4). Generally, researchers have observed that the amount of crop residue can be estimated from grain yield. For example, linear relationships between the amount of crop residue and grain yield have been reported for small grains (McCool et al., 2006) and corn (Linden et al., 2000), such that grain yield has been used to predict residue yield (Johnson et al., 2006a).

The relationship between residue amount and groundcover varies with crop. Gregory (1982) presents a method to estimate the fraction of groundcover from the mass of residue per area of ground, using coefficients determined from field studies. For each crop, Gregory reported that the fraction of soil covered increased exponentially with increasing residue amount, with the lowest rate of increase for cotton (*Gossypium* sp.) and the highest rate of increase for oats (*Avena sativa* L.). Thus, the amount of soil cover generally increases exponentially with grain yield. This exponential relationship indicates that harvesting residue will not result in a proportional decrease in groundcover. For example, under most circumstances, harvesting 25% of the crop residue mass will decrease the amount of groundcover by less than 25%.

Roots and growing plant materials also effectively reduce wind and water erosion and show trends similar to crop residues. Soil loss rates by water erosion decrease exponentially with increasing vegetative cover; soil loss is approximately the same for 60% and 100% vegetation cover (Stocking, 1988). Research also has shown exponential decreases in soil loss rates by water erosion with increasing root mass (Gyssels et al., 2005).

Cover crops and perennials also may reduce soil erosion drastically by keeping the soil covered. The amount of soil cover provided by growing plants, their height, structure, orientation, rooting characteristics, position, and other factors are important in determining their effectiveness at reducing erosion. For example, erosion from land planted to woody species can be comparable to traditional row-cropped land if no cover crop is present (Malik et al., 2000), and erosion rates can be very high in mature forests (Stocking, 1988). Positioning of plants plays a role in their ability to reduce erosion. In a field study, measured soil loss exceeded the predicted wind erosion soil loss because plants were sited between soil ridges, limiting their ability to reduce erosion (Van Donk and Skidmore, 2003).

For residue to be effective for decreasing erosion, it must cover the soil surface during the erosive event. Thus, residue must remain on the soil surface until the next crop is established. Tillage, seeding, and other soil disruptions decrease crop residue present on soil surface. Extensive research has been conducted to determine soil erosion rates under differing conditions of cover following tillage. Eck et al. (2001) reported that each tillage operation can decrease the crop residue cover by 10% to 20% (for mild disturbance caused by some drills or planters) to 95% or more (for aggressive tillage such as a moldboard plow). Tillage incorporates residue into the soil, where it can still contribute to C cycling and nutrient cycling, but harvesting residue removes soil cover along with C and other nutrients.

Soils with poor aggregate structures exhibit less resilience against erosive forces such as wind and water. In both tilled and not-tilled soils, residue harvest increased the number of small aggregates susceptible to wind erosion (Singh et al., 1994; Malhi et al., 2006; Singh and Malhi, 2006; Malhi and Kutcher, 2007; Malhi and Lemke, 2007). Blanco-Canqui and Lal (2007) also reported that removing wheat straw reduced soil aggregate strength compared to mulching. Tillage can increase erosion through decreased aggregate stability and increased soil detachment. Conversely, under some conditions, tillage can decrease soil loss by wind and water due to increased water infiltration and soil surface roughness (Dabney et al., 2004).

The effect of interaction between tillage and crop residue on soil loss by erosion is complex and varies with soil properties such as moisture (Cogo et al., 1982; Dabney et al., 2004). Reduced tillage can provide more soil cover (Guy and Cox, 2002), but residue removal can negate some of the benefits of reduced tillage. In both the northern and southern U.S., removing corn residue from reduced-till or no-till plots can result in soil loss rates by water erosion similar to those for conventionally tilled soil with no residue removed (Lindstrom, 1986; Dabney et al., 2004). In no-till soils, the portion of standing residue relative to flat residue changes with time (Steiner et al., 2000), and this is expected to alter the effectiveness of remaining residues in reducing wind and water erosion. McCool et al. (2006) noted that for small grains, stems are the most important components for reducing erosion because they are more resistant to degradation and relocation than leaves. Some studies suggest that decomposition of corn residue over winter reduces cover by 20% to 30% (Van Donk and Skidmore, 2003; Wilson et al., 2008). Residue decomposition rates vary with the chemical composition of plant materials, temperature, moisture, soil characteristics, and placement (Paul, 1991).

The highly non-linear nature of the erosion process and the simultaneous interactions of numerous processes make soil loss by erosion very difficult to predict. The amount of soil lost by erosion is a complex function of macro- and microtopography, the energy of the wind or water impacting the soil, erodibility, and other factors. Current models based on years of research indicate soil cover is a critical determinant of soil loss by wind and water erosion (Figure 1.4) that varies with soil types (Lindstrom, 1986; Erenstein, 2002). Additional research is needed to more completely characterize the implications of biomass harvest on long-term and episodic soil erosion in relation to biofuel production.

WATERSHED HYDROLOGICAL IMPACTS

Generally, much of the preceding material is based on studies of residue cover as affected by tillage practices (i.e., incorporation) rather than removal. Most of the work was done at plot scale. Extending this knowledge to include effects on watershed hydrology is difficult. Uhlenbrook (2007) stated that no research had been published on the impacts of biofuel development on watershed hydrology. It is critical to develop an understanding of these impacts as soon as possible and more clearly appreciate the differences between residue incorporation and removal in terms of their relative effects on interacting C, nutrient, and water cycles.

As more land becomes dedicated to producing biofuel crops, environmental impacts of associated land use conversions will depend on the nature and extent of changes in land cover and vegetation management. Any change in land use will influence the partitioning of precipitation into canopy interception, overland flow, evaporation, transpiration, and deep percolation, along with accompanying hydrological consequences. In the tropics, land use may shift toward clearing of forests and expansion of agricultural areas with hydrological consequences that are difficult to model due to limited datasets covering hydrology of tropical watersheds (Uhlenbrook, 2007).

In temperate zones, land use conversion for biofuel crops may expand perennial cover at the expense of annual crop cover. Short-rotation tree crops (*Poplar* or *Salix*) or tall prairie grass species (switchgrass) can be highly productive in semi-arid to humid temperate climates, and probably require fewer nutrient inputs than annual crops (Johnson et al., 2007d). Land cover conversions to these perennial crops would increase transpiration and reduce overland flow (Rachman et al., 2004; Updegraffa et al., 2004), and have been shown to sequester more soil C than corn in highly fertile soils (Zan et al., 2001). This would benefit the hydrologic regimens of Midwestern streams and rivers based on recent trends of increasing precipitation amounts and intensities that are predicted to continue (Nearing et al., 2004; Hodgkins et al., 2007).

To the extent that bioenergy feedstocks are derived from residues of annual crops such as corn, potential hydrologic impacts may lean toward greater fractions of precipitation lost via overland flow and less near-surface soil moisture (Rhoton et al., 2002; Montgomery, 2007). Tillage practices that incorporate residue were shown to increase the overland flow component of stream discharge from small watersheds by nearly 50% in one long-term (25-year) study (Tomer et al., 2005). The differences occurred during large rainfall events (up to 100 mm d^{-1}) and intermediate and small runoff-producing events. At the same time, watersheds with conservation tillage systems like ridge tills yielded greater baseflow and total discharge, and showed more rapid recovery from drought. The increase in discharge was about 3% of the total hydrologic budget, accompanied by less variation in streamflow. Conservation tillage resulted in lower bulk densities, greater SOM, and under wet soil conditions, greater water contents than conventional tillage (Tomer et al., 2006). Differences in hydrology likely result from the direct impacts of soil cover on water flow and changes in aggregate stability and infiltration capacity. Stover removal reduced saturated conductivity on three Ohio soils (Blanco-Canqui et al., 2007), and is expected to result in greater overland flow and decreased baseflow. Ensuring adequate ground cover following residue harvest may be critical to increase infiltration and reduce surface runoff, especially if climate change increases amounts and intensities of precipitation.

INTEGRATION OF CARBON, NUTRIENT, AND WATER CYCLES

Removal of non-grain biomass simultaneously interacts with C, nutrient, microclimate, and hydrological cycles (Figure 1.5). Harvesting residue in excess of MSC will reduce SOC. Excess harvesting limits the organic material needed for soil aggregation, making soil more susceptible to erosive forces. Removal of biomass can lead to surface sealing of soil, reducing infiltration and increasing surface runoff. Surface runoff across unprotected soils removes top soil and the nutrients it contains. An influx of P and K into surface water promotes algal blooms, eutrophication, and hypoxia (Kim and Dale, 2005). Over time, soil erosion can result in exposed subsoil that typically is less fertile with less SOM

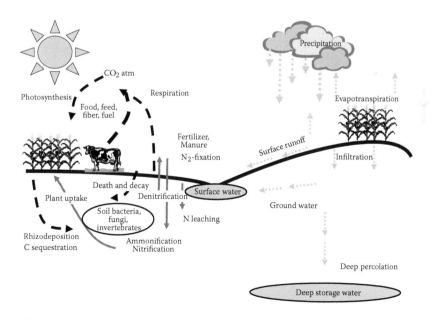

FIGURE 1.5 Interconnection of carbon, nitrogen, and water cycles.

compared to surface soil, thus impeding soil nutrient cycling and decreasing nutrient-holding capacity. As a result, additional nutrients are required to support production.

The lack of biomass inputs decreases soil fauna (Tables 1.3 through 1.5) and can interfere with nutrient cycling. Retaining biomass on the surface can promote infiltration, reducing surface runoff, but can increase the potential for leaching nutrients such as nitrates into ground water. Surface residue keeps the soil surface cooler, slowing evaporation and promoting denitrifying conditions by extending the duration of anaerobic soil conditions (Ball et al., 1999; Aulakh et al., 2001). Cooler surface soil can delay germination and retard early season growth in regions with cool, wet springs (Swan et al., 1994). In warmer drier climates, the lack of surface cover promotes water stress and decreases yield (Power et al., 1986; Wilhelm et al., 1986). The interactions of residue management with biological, chemical and physical processes are complicated by climatic factors and management practices. The key is finding a balanced non-grain biomass harvest approach that supports soil processes and controls erosion to minimize potential negative effects of non-grain biomass removal.

BIOMASS HARVEST: COMPENSATION STRATEGIES

Several strategies can avoid or reduce loss of SOM and solve related problems arising from biomass harvest. Harvest rates should be limited to those that maintain SOM and do not exacerbate erosion. Reducing or eliminating tillage utilizes remaining residue as ground cover to reduce erosion. If harvest rates exceed the amount needed to provide adequate inputs for SOM, alternative inputs such as manure should be applied. In general, manures tend to increase SOC under a wide range of management and climatic conditions (Johnson et al., 2007b). Animal manures contain 40% to 60% C on a dry weight basis and can promote SOC sequestration and provide nutrient inputs (CAST, 1992). Another strategy is planting cover crops and living mulches where crop residues are harvested to prevent erosion and replace C and N removed the residues (Zemenchik et al., 2000; Drinkwater and Snapp, 2007).

Other amendments such as application of by-products of cellulosic fermentation containing high lignin concentrations improved soil quality characteristics in laboratory studies (Johnson et al., 2004; Johnson et al., 2007c). Another by-product is biochar from pyrolysis or gasification. Biochar has the potential to enhance plant growth by supplying and retaining nutrients and improving soil physical and chemical properties. Biochar may also remove pesticides or other pollutants from soil water (Glaser et al., 2002; Lehmann et al., 2003; Lehmann et al., 2006; Lehmann and Rondon, 2006). Compensation strategies will vary by management system, climatic regime, and suitability of strategy to farming systems.

SUMMARY

Harvesting crop non-grain biomass initiates a cascade of interrelated biological, chemical, and physical soil events. Biomass harvest has the potential to disrupt soil nutrient dynamics, water relations, and other important soil processes. Considerable knowledge exists about ways to minimize the risks of harvesting non-grain biomass.

It is essential that regional and site-specific guidelines be developed as quickly as possible. Clearly, non-grain biomass harvest must be limited to avoid loss of SOM and prevent excessive soil erosion.

Management strategies that enhance soil quality such as reduced or no tillage, inclusion of perennial species, use of cover crops and living mulches, and applying amendments such as biochar or manure may compensate for non-grain biomass removal. Soil quality building strategies reduce the risk of erosion, improve C and nutrient cycling, and increase aggregate stabilization. In developing a biofuel economy, it is paramount that soil resources be protected to secure our nation's ability to provide adequate food, feed, fiber, and fuel for a growing world.

ACKNOWLEDGMENTS

This project is supported through the Renewable Energy Assessment Project (REAP) of the United States Department of Agriculture's Agricultural Research Service* and also by the Sun Grant Initiative of the United States Department of Energy. The authors thank K. Eystad for assistance with Figure 1.5 and B. Burmeister for her careful editing.

REFERENCES

Abiven, S. et al. 2005. Mineralisation of C and N from root, stem and leaf residues in soil and role of their biochemical quality. *Biol. Fertil. Soils* 42: 119–128.

Adler, P.R. et al. 2006. Biomass yield and biofuel quality of switchgrass harvested in fall or spring. *Agron. J.* 98: 1518–1525.

Al-Kaisi, M.M., X. Yin, and M.A. Licht. 2005. Soil carbon and nitrogen changes as affected by tillage system and crop biomass in a corn–soybean rotation. *Appl. Soil Ecol.* 30: 174–191.

Al-Yemeni, M.N. and J. Grace. 1995. Radiation balance of an alfalfa crop in Saudi Arabia. *J. Arid Environ.* 29: 447–454.

Albright, W.A. 1938. Loss of soil organic matter and its restoration. In *Soils and Men: Yearbook of Agriculture*. Washington, D.C., U.S. Department of Agriculture, pp. 347–360.

Allmaras, R.R., D.R. Linden, and C.E. Clapp. 2004. Corn residue transformations into root and soil carbon as related to nitrogen, tillage, and stover management. *Soil Sci. Soc. Am. J.* 68: 1366–1375.

Amos, B. and D.T. Walters. 2006. Maize root biomass and net rhizodeposited carbon: An analysis of the literature. *Soil Sci. Soc. Am. J.* 70: 1489–1503.

Andren, O. and K. Paustian. 1987. Barley straw decomposition in the field: A comparison of models. *Ecology* 68: 163–210.

Angers, D.A. 1998. Water-stable aggregation in Quebec silty clay soil: Some factors controlling its dynamics. *Soil Tillage Res.* 47: 91–96.

Angers, D.A., A. Pesant, and J. Vigneux. 1992. Early cropping-induced changes in soil aggregation, organic matter, and microbial biomass. *Soil Sci. Soc. Am. J.* 56: 115–119.

Aulakh, M.S. et al. 2001. Denitrification, N_2O and CO_2 fluxes in rice–wheat cropping system as affected by crop residues, fertilizer N and legume green manure. *Biol. Fertil. Soils* 34: 675–689.

* The United States Department of Agriculture is an equal opportunity provider and employer.

Bainbridge, D.A. 1986. High performance low cost buildings of straw. *Agric. Ecosyst. Environ.* 16: 281–284.

Balesdent, J. and M. Balabane. 1996. Major contribution of roots to soil carbon storage inferred from maize cultivated soils. *Soil Biol. Biochem.* 28: 1261–1263.

Ball, B.C., J.P. Parker, and A. Scott. 1999. Soil and residue management effects on cropping conditions and nitrous oxide fluxes under controlled traffic in Scotland 2. Nitrous oxide, soil N status and weather. *Soil Tillage Res.* 52: 191–201.

Barber, S.A. 1979. Corn residue management and soil organic matter. *Agron. J.* 71: 625–627.

Bayer, C. et al. 2006. A method for estimating coefficients of soil organic matter dynamics based on long-term experiments. *Soil Tillage Res.* 91: 217–226.

Beare, M.H. and R.R. Bruce. 1993. A comparison of methods for measuring water-stable aggregates: Implications for determining environmental effects on soil structure. *Geoderma* 56: 87–104.

Bending, G.D., M.K. Turner, and J.E. Jones. 2002. Interactions between crop residue and soil organic matter quality and the functional diversity of soil microbial communities. *Soil Biol. Biochem.* 34: 1073–1082.

Benjamin, J.G., M.M. Mikha, and M.F. Vigil. 2008. Organic carbon effects on soil physical and hydraulic properties in a semiarid climate. *Soil Sci. Soc. Am. J.* 72: 1357–1362.

Benoit, G.R. et al. 1986. Tillage-residue effects on snow cover, soil water, temperature and frost. *Trans. ASAE* 29: 473–479.

Bhatnagar, V.K., T.N. Chaudhary, and B.R. Sharma. 1983. Effect of tillage and residue management on properties of two coarse-textured soils and on yield of irrigated wheat and groundnut. *Soil Tillage Res.* 3: 27–37.

Bilbro, J.D. and D.W. Fryrear. 1994. Wind erosion losses as related to plant silhouette and soil cover. *Agron. J.* 86: 550–553.

Black, A.L. 1973. Soil property changes associated with crop residue management in a wheat–fallow rotation. *Soil Sci. Soc. Am. J.* 37: 943–946.

Blanco-Canqui, H. and R. Lal. 2007. Soil structure and organic carbon relationships following 10 years of wheat straw management in no-till. *Soil Tillage Res.* 95: 240–254.

Blanco-Canqui, H. et al. 2006a. Changes in long-term no-till corn growth and yield under different rates of stover mulch. *Agron. J.* 98: 1128–1136.

Blanco-Canqui, H. et al. 2006b. Corn stover impacts on near-surface soil properties of no-till corn in Ohio. *Soil Sci. Soc. Am. J.* 70: 266–278.

Blanco-Canqui, H. et al. 2007. Soil hydraulic properties influenced by corn stover removal from no-till corn in Ohio. *Soil Tillage Res.* 92: 144–155.

Blevins, R.L. and W.W. Frye. 1993. Conservation tillage: An ecological approach to soil management. *Adv. Agron.* 51: 33–78.

Borie, F. et al. 2002. Interactions between crop residues application and mycorrhizal developments and some soil-root interface properties and mineral acquisition by plants in an acidic soil. *Biol. Fertil. Soils* 36: 151–160.

Bransby, D.I., S.B. McLaughlin, and D.J. Parrish. 1998. A review of carbon and nitrogen balances in switchgrass grown for energy. *Biom. Bioenergy* 14: 379–384.

Breakwell, D.P., and R.F. Turco. 1989. Nutrient and phytotoxic contributions of residue to soil in no-till continuous-corn ecosystems. *Biol. Fertil. Soils* 8: 328–334.

Bristow, K.L. 1988. The role of mulch and its architecture in modifying soil temperature. *Aust. J. Soil Res* 26: 269–280.

Bristow, K.L. et al. 1986. Simulation of heat and moisture transfer through a surface residue soil system. *Agric. For. Meteorol.* 36: 193–214.

Buerkert, A., A. Bationo, and K. Dossa. 2000. Mechanisms of residue mulch-induced cereal growth increases in West Africa. *Soil Sci. Soc. Am. J.* 64: 346–358.

Burgess, M.S., G.R. Mehuys, and C.A. Madramootoo. 2002. Decomposition of grain corn residues (*Zea mays* L.): Litterbag study under three tillage systems. *Can. J. Soil Sci.* 82: 127–138.

Bussière, F. and P. Cellier. 1994. Modification of the soil temperature and water content regimes by a crop residue mulch: Experiment and modeling. *Agric. For. Meteorol.* 68: 1–28.

Campbell, C.A. et al. 1991. Effect of crop rotations and fertilization on soil organic matter and some biochemical properties of a thick black chernozem. *Can. J. Soil Sci.* 71: 377–387.

Cassida, K.A. et al. 2005. Biofuel component concentrations and yields of switchgrass in South Central U.S. environments. *Crop Sci.* 45: 682–692.

CAST. 1992. Preparing U.S. agriculture for global climate change. Task Force Report 119. Council for Agricultural Science and Technology, Ames, Iowa.

Chaney, K. and R.S. Swift. 1984. The influence of organic matter on aggregate stability in some British soils. *J. Soil Sci.* 35: 223–230.

Christensen, B.T. 1986. Barley straw decomposition under field conditions: Effects of placement and initial nitrogen content on weight loss and nitrogen dynamics. *Soil Biol. Biochem.* 18: 523–529.

Chung, S.O. and R. Horton. 1987. Soil heat and water flow with a partial surface mulch. *Water Resour. Res.* 23: 2175–2186.

Clay, D.E. et al. 2006. Theoretical derivation of stable and nonisotopic approaches for assessing soil organic carbon turnover. *Agron. J.* 98: 443–450.

Clay, D.E. et al. 2001. Factors influencing spatial variability of apparent electrical conductivity. *Commun. Soil Sci. Plt. Anal.* 32: 2993–3008.

Clifton-Brown, J.C. and I. Lewandowski. 2002. Screening miscanthus genotypes in field trials to optimise biomass yield and quality in Southern Germany. *Eur. J. Agron.* 16: 97–110.

Cogo, N.P., W.C. Moldenhauer, and G.R. Foster. 1982. Soil loss reductions from conservation tillage practices. *Soil Sci. Soc. Am. J.* 48: 368–373.

Cook, R.J., M.G. Boosalis, and B. Doupnik. 1978. Influence of crop residue on plant diseases. In Oschwald, W.R. et al., Eds., *Crop Residue Management Systems*, Special Publication 31. SSSA, Madison, WI, pp. 147–163.

Cookson, W.R., M.H. Beare, and P.E. Wilson. 1998. Effects of prior crop residue management on microbial properties and crop residue decomposition. *Appl. Soil Ecol.* 7: 179–188.

Cox, W.J. et al. 1990. Tillage effects on some soil physical and corn physiological characteristics. *Agron. J.* 82: 806–812.

Crookston, R.K. et al. 1991. Rotational cropping sequence affects yield of corn and soybeans. *Agron. J.* 83: 108–113.

Dabney, S.M. et al. 2004. History, residue, and tillage effects on erosion of loessial soil. *Trans. ASAE* 47: 767–775.

Dam, R.F. et al. 2005. Soil bulk density and crop yield under eleven consecutive years of corn with different tillage and residue practices in a sandy loam soil in central Canada. *Soil Tillage Res.* 84: 41–53.

De Gryze, S., J. Six, C. Brits, and R. Merckx. 2005. A quantification of short-term macroaggregate dynamics: Influences of wheat residue input and texture. *Soil Biol. Biochem.* 37: 55–66.

Debosz, K., P.H. Rasmussen, and A.R. Pedersen. 1999. Temporal variations in microbial biomass C and cellulolytic enzyme activity in arable soils: Effects of organic matter input. *Appl. Soil Ecol.* 13: 209–218.

Degens, B.P. 1997. Macroaggregation of soil by biological bonding and binding mechanisms and the factors affecting these: A review. *Aust. J. Soil Res.* 35: 431–459.

Dexter, A.R., D. Hein, and J.S. Hewett. 1982. Macrostructure of the surface layer of self-mulching clay in relation to cereal stubble management. *Soil Tillage Res.* 2: 251–264.

Dhima, K.V. et al. 2006. Allelopathic potential of winter cereals and their cover crop mulch effect on grass weed suppression and corn development. *Crop Sci.* 46: 345–352.

Diaz-Zorita, M. and G.A. Grosso. 2000. Effect of soil texture, organic carbon and water retention on the compactability of soils from the Argentinean pampas. *Soil Tillage Res.* 54: 121–126.

Dipardo, J. 2000. Outlook for biomass ethanol production and demand. U.S. Energy Information Agency, Washington, D.C. http://www.eia.doe.gov/oiaf/analysispaper/pdf/biomass.pdf (posted March 10, 2002; verified Sept. 4, 2008).

Doran, J.W. and T.B. Parkin. 1994. Defining and assessing soil quality. In Doran, J.W. et al., Eds., *Defining Soil Quality for a Sustainable Environment,* Special Publication 35. SSSA, Madison, WI, pp. 3–21.

Dormaar, J.R. 1983. Chemical properties of soil and water-stable aggregation after 67 years of cropping to spring wheat. *Plant Soil* 75: 51–61.

Drinkwater, L.E. and S.S. Snapp. 2007. Nutrients in agroecosystems: Rethinking the management paradigm. *Adv. Agron.* 92: 63–186.

Drury, C.F. et al. 2003. Impacts of zone tillage and red clover on corn performance and soil physical quality. *Soil Sci. Soc. Am. J.* 67: 867–877.

Duffy, M.D. and V.Y. Nanhou. 2001. Costs of producing switchgrass for biomass in Iowa. <http://www.extension.iastate.edu/Publications/PM1866.pdf> (posted April 2001; verified Sept. 5, 2008).

Duley, F.L. and J.C. Russel. 1939. The use of crop residues for soil and moisture conservation. *J. Am. Soc. Agron.* 31: 703–709.

Duppong, L.M. et al. 2004. The effect of natural mulches on crop performance, weed suppression and biochemical constituents of catnip and St. John's wort. *Crop Sci.* 44: 861–869.

Eck, K.J., D.E. Brown, and A.B. Brown. 2001. Estimating corn and soybean residue cover. Purdue University Cooperative Extension Service. http://www.agry.purdue.edu/ext/pubs/AY-269-W.pdf (verified Sept. 4, 2008).

Edwards, A.P. and J.M. Bremner. 1967. Microaggregates in soils. *J. Soil Sci.* 18: 64–73.

Edwards, W.M. et al. 1993. Factors affecting preferential flow of water and atrazine through earthworm burrows under continuous no-tillage corn. *J. Environ. Qual.* 22: 453–457.

Edwards, W.M. et al. 1990. Effect of *Lumbricus terrestris* L. burrows on hydrology of continuous no-till corn fields. *Geoderma* 46: 73–84.

Ekenler, M. and M.A. Tabatabai. 2003. Tillage and residue management effects on β-glucosaminidase activity in soils. *Soil Biol. Biochem.* 35: 871–874.

Elliott, E.T. 1986. Aggregate structure and carbon, nitrogen, and phosphorus in native and cultivated soils. *Soil Sci. Soc. Am. J.* 50: 627–633.

Elliott, E.T. and D.C. Coleman. 1988. Let the soil work for us. *Ecol. Bull* 39: 23–32.

Enz, J.W., L.J. Brun, and J.K. Larsen. 1988. Evaporation and energy balance for bare and stubble covered soil *Agric. Forest Meteor.* 43: 59–70.

Erenstein, O. 2002. Crop residue mulching in tropical and semi-tropical countries: Evaluation of residue availability and other technological implications. *Soil Tillage Res.* 67: 115–133.

Fageria, N.K. 2004. Dry matter yield and shoot nutrient concentrations of upland rice, common bean, corn, and soybean grown in rotation on an oxisol. *Commun. Soil Sci. Plant Anal.* 35: 961–974.

FAO. 2004. Unified bioenergy terminology. Food and Agriculture Organization of the United Nations, Forestry Department, Wood Energy Programme. ftp://ftp.fao.org/docrep/fao/007/j4504e/j4504e00.pdf (posted 2004; verified Sept. 4, 2008).

Fatondji, D. et al. 2006. Effect of planting technique and amendment type on pearl millet yield, nutrient uptake, and water use on degraded land in Niger. *Nutr. Cycl. Agroecosys.* 76: 203–217.

Follett, R.F., J.Z. Castellanos, and E.D. Buenger. 2005. Carbon dynamics and sequestration in an irrigated vertisol in Central Mexico. *Soil Tillage Res.* 83: 148–158.

Ford, J.H. and D.R. Hicks. 1992. Corn growth and yield in uneven emerging stands. *J. Prod. Agric.* 5: 185–188.

Franck, W.M. et al. 1997. Decomposition of litter produced under elevated CO_2: Dependence on plant species and nutrient supply. *Biogeochemistry* 36: 223–237.

Frank, A.B. et al. 2004. Biomass and carbon partitioning in switchgrass. *Crop Sci.* 44: 1391–1396.

Franzluebbers, A.J., F.M. Hons, and V.A. Saladino. 1995. Sorghum, wheat and soybean production as affected by long-term tillage, crop sequence and N fertilization. *Plant Soil* 173: 55–65.

Fraser, P.M. and J.E. Piercy. 1998. The effects of cereal straw management practices on lumbricid earthworm populations. *Appl. Soil Ecol.* 9: 369–373.

Fraser, P.M., P.H. Williams, and R.J. Haynes. 1996. Earthworm species, population size and biomass under different cropping systems across the Canterbury Plains, New Zealand. *Appl. Soil Ecol.* 3: 49–57.

Frey, S.D., E.T. Elliott, and K. Paustian. 1999. Bacterial and fungal abundance and biomass in conventional and no-tillage agroecosystems along two climatic gradients. *Soil Biol. Biochem.* 31: 573–585.

Garcia-Huidobro, J., J. Monteith, and G. Squire. 1982. Time, temperature, and germination of pearl millet (*Pennisetum typhoides* S&H). *J. Exp. Bot.* 33: 288–296.

Gäth, S. and H.G. Frede. 1995. Mechanisms of air slaking. In Hartge, K.H. and Stewart, B.A., Eds., *Soil Structure: Its Development and Function*. Advances in Soil Science Series. CRC Press, Boca Raton, FL, pp. 159–173.

Gercel, H.F. 2002. Production and characterization of pyrolysis liquids from sunflower-pressed bagasse. *Bioresour. Technol.* 85: 113–117.

Ghidey, F. and E. Alberts. 1993. Residue type and placement effects on decomposition: Field study at model evaluation. *Trans. ASAE* 36: 1611–1617.

Glaser, B., J. Lehman, and W. Zech. 2002. Ameliorating physical and chemical properties of highly weathered soils in the tropics with charcoal: Review. *Biol. Fertil. Soils* 35: 219–230.

Golchin, A., J.M. Oades, J.O. Skjemstad, and P. Clark. 1994a. Soil structure and carbon cycling. *Aust. J. Soil Res.* 32: 1043–1068.

Golchin, A. et al. 1994b. Study of free and occluded particulate organic matter in soils by solid state ^{13}C CP/MAS NMR spectroscopy and scanning electron microscopy. *Aust. J. Agric. Res.* 32: 285–309.

Govaerts, B. et al. 2007a. Infiltration, soil moisture, root rot and nematode populations after 12 years of different tillage, residue and crop rotation managements. *Soil Tillage Res.* 94: 209–219.

Govaerts, B. et al. 2007b. Influence of tillage, residue management, and crop rotation on soil microbial biomass and catabolic diversity. *Appl. Soil Ecol.* 37: 18–30.

Govaerts, B. et al. 2008. Long-term consequences of tillage, residue management, and crop rotation on selected soil micro-flora groups in the subtropical highlands. *Appl. Soil Ecol.* 38: 197–210.

Graham, R.L. et al. 2007. Current and potential U.S. corn stover supplies. *Agron. J.* 99: 1–11.

Gregory, J.M. 1982. Soil cover prediction with various amounts and types of crop residue. *Trans. ASAE* 25: 1333–1337.

Guérif, J. 1990. Factors influencing compaction-induced increases in soil strength. *Soil Tillage Res.* 16: 167–178.

Gupta, V.V.S.R. and J.J. Germida. 1988. Distribution of microbial biomass and its activity in different soil aggregate size classes as affected by cultivation. *Soil Biol. Biochem.* 20: 777–786.

Guy, S.O. and D.B. Cox. 2002. Reduced tillage increases residue groundcover in subsequent dry pea and winter wheat crops in the Palouse region of Idaho. *Soil Tillage Res.* 66: 69–77.

Gyssels, G. et al. 2005. Impact of plant roots on the resistance of soil to erosion by water: Review. *Progr. Phys. Geogr.* 29: 189–221.

Hagen, L.J. 1996. Crop residue effects on aerodynamic processes and wind erosion *J. Theor. Appl. Clim.* 54: 39–46.

Halvorson, A.D. and J.M.F. Johnson. 2009. Irrigated corn cob production in the Central Great Plains. *Agron. J.*, in press.

Halvorson, A.D. and C.A. Reule. 2007. Irrigated, no-till corn and barley response to nitrogen in Northern Colorado. *Agron. J.* 99: 1521–1529.

Havlin, J.L. and D.E. Kissel. 1997. Management effects on soil organic carbon and nitrogen in the east-central Great Plains of Kansas. In Paul, E.A. et al., Eds., *Soil Organic Matter in Temperate Agroecosystems: Long-Term Experiments in North America.* CRC Press, Boca Raton, FL, pp. 381–386.

Haynes, R.J., R.S. Swift, and R.C. Stephen. 1991. Influence of mixed cropping rotations (pasture-arable) on organic matter content, stable aggregation and clod porosity in a group of soils. *Soil Tillage Res.* 19: 77–87.

Heal, O.W., J.M. Anderson, and M.J. Swift. 1997. Plant litter quality and decomposition: Historical overview. In Cadish, G. and Giller, K.E., Eds., *Driven by Nature: Plant Litter Quality and Decomposition.* CAB International, Wallingford, U.K., pp. 47–66.

Hettenhaus, J.R., R. Wooley, and A. Wiselogel. 2000. Biomass commercialization prospects in the next 2 to 5 years: Biomass colloquies 2000 NREL/ACO-9-29-039-01. National Renewable Energy Laboratory, Golden, CO. http://www.nrel.gov/docs/fy01osti/28886.pdf (posted Oct. 10, 2000; verified March 1, 2007).

Hillel, D. 1998. *Environmental Soil Physics.* Academic Press, San Diego, CA.

Hobbs, J.A. and P.I. Brown. 1965. Effects of cropping and management on nitrogen and organic carbon contents of a western Kansas soil. *Agric. Exp. Stn. Tech. Bull.* 144. Kansas State University, Manhattan.

Hodgkins, G.A., R.W. Dudley, and S.S. Aichele, Eds. 2007. Historical changes in precipitation and streamflow in the U.S. Great Lakes Basin. 1915–2004. U.S. Geological Survey Scientific Investigations Report 5118. Reston, VA, p. 37.

Hooker, B.A. et al. 2005. Long-term effects of tillage and corn stalk return on soil carbon dynamics. *Soil Sci. Soc. Am. J.* 69: 188–196.

Horner, G.M. et al. 1960. Effect of cropping practices on yield, soil organic matter, and erosion in the Pacific Northwest wheat region. USDA ARS Coop. Bull. 1, Pullman, WA.

Horton, R. et al. 1996. Crop residue effects on surface radiation and energy balance: Review. *Theor. Appl. Clim.* 54: 27–37.

Hoskinson, R.L. et al. 2007. Engineering, nutrient removal, and feedstock conversion evaluations of four corn stover harvest scenarios. *Biom. Bioenergy* 31: 126–136.

Huggins, D.R. et al. 2007. Corn-soybean sequence and tillage effects on soil carbon dynamics and storage. *Soil Sci. Soc. Am. J.* 71: 145–154.

Huggins, D.R. et al. 1998. Carbon dynamics in corn–soybean sequences as estimated from natural carbon-13 abundance. *Soil Sci. Soc. Am. J.* 62: 195–203.

Islam, M.N. and F.N. Ani. 2000. Techno-economics of rice husk pyrolysis, conversion with catalytic treatment to produce liquid fuel. *Bioresourc. Technol.* 73: 67–75.

Jalota, S.K., R. Khera, and S.S. Chahal. 2001. Straw management and tillage effects on soil water storage under field conditions. *Soil Use Mgt.* 17: 282–287.

Jastrow, J.D. 1996. Soil aggregate formation and the accrual of particulate and mineral-associated organic matter. *Soil Biol. Biochem.* 28: 656–676.

Jastrow, J.D. and R.M. Miller. 1997. Soil aggregate stabilization and carbon sequestration: Feedbacks through organomineral associations. In Lal, R. et al., Eds., *Soil Processes and the Carbon Cycle.* CRC Press, Boca Raton, FL, pp. 207–223.

Jawson, M.D. and L.F. Elliott. 1986. Carbon and nitrogen transformation wheat straw and root decomposition. *Soil Biol. Biochem.* 18: 15–22.

Jiang, P. and K.D. Thelen. 2004. Effect of soil and topographic properties on crop yield in a north central corn–soybean cropping system. *Agron. J.* 96: 252–258.

Johnson, J.M.F., R.R. Allmaras, and D.C. Reicosky. 2006a. Estimating source carbon from crop residues, roots and rhizodeposits using the national grain-yield database. *Agron. J.* 98: 622–636.

Johnson, J.M.F. et al. 2006b. A matter of balance: Conservation and renewable energy. *J. Soil Water Conserv.* 61: 120A–125A.

Johnson, J.M.F., N.W. Barbour, and S.L. Weyers. 2007a. Chemical composition of crop biomass impacts its decomposition. *Soil Sci. Soc. Am. J.* 71: 155–162.

Johnson, J.M.F. et al. 2007b. Agricultural opportunities to mitigate greenhouse gas emissions. *Environ. Pollut.* 150: 107–124.

Johnson, J.M.F. et al. 2007c. Impact of high lignin fermentation by-product on soils with contrasting soil organic carbon. *Soil Sci. Soc. Am. J.* 71: 1151–1159.

Johnson, J.M.F. et al. 2007d. Biomass-bioenergy crops in the United States: Changing paradigm. *Am. J. Plant Sci. Biotechnol.* 1: 1–28.

Johnson, J.M.F. et al. 2004. Characterization of soil amended with the by-product of corn stover fermentation. *Soil Sci. Soc. Am. J.* 68: 139–147.

Johnson, P.A. and B.J. Chamber. 1996. Effects of husbandry on soil organic matter. *Soil Use Mgt.* 13: 102–103.

Kaewpradit, W. et al. 2008. Regulating mineral N release and greenhouse gas emissions by mixing groundnut residues and rice straw under field conditions. *Eur. J. Soil Sci.* 59: 640–652.

Kapkiyai, J.J. et al. 1999. Soil organic matter and nutrient dynamics in a Kenyan nitisol under long-term fertilizer and organic input management. *Soil Biol. Biochem.* 31: 1773–1782.

Karlen, D.L. et al. 1994. Crop residue effects on soil quality following 10 years of no-till corn. *Soil Tillage Res.* 31: 149–167.

Kato, Y. et al. 2007. Improvement of rice (*Oryza sativa* L.) growth in upland conditions with deep tillage and mulch. *Soil Tillage Res.* 92: 30–44.

Katterer, T., O. Andren, and R. Pettersson. 1998. Growth and nitrogen dynamics of reed canary grass (*Phalaris arundinacea* L.) subjected to daily fertilization and irrigation in the field. *Field Crops Res.* 55: 153–164.

Kim, S. and B.E. Dale. 2005. Life cycle assessment of various cropping systems utilized for producing biofuels: Bioethanol and biodiesel. *Biom. Bioenergy* 29: 426–439.

Kladivko, E.J. 2001. Tillage systems and soil ecology. *Soil Tillage Res.* 61: 61–76.

Kong, A.Y.Y. et al. 2005. The relationship between carbon input, aggregation, and soil organic carbon stabilization in sustainable cropping systems. *Soil Sci. Soc. Am. J.* 69: 1078–1085.

Krzic, M. et al. 2004. Soil properties influencing compactability of forest soils in British Columbia. *Can. J. Soil Sci.* 84: 219–226.

Kucharik, C.J. et al. 2001. Measurements and modeling of carbon and nitrogen cycling in agroecosystems of southern Wisconsin: Potential for SOC sequestration during the next 50 years. *Ecosystems* 4: 237–258.

Kumar, K. and K.M. Goh. 2000. Crop residues and management practices: effects on soil quality, soil nitrogen dynamics, crop yield and nitrogen recovery. *Adv. Agron.* 68: 197–219.

Kundu, S. et al. 2007. Carbon sequestration and relationship between carbon addition and storage under rainfed soybean–wheat rotation in a sandy loam soil of the Indian Himalayas. *Soil Tillage Res.* 92: 87–95.

Kushwaha, C.P., S.K. Tripathi, and K.P. Singh. 2000. Variations in soil microbial biomass and N availability due to residue and tillage management in a dryland rice agroecosystem. *Soil Tillage Res.* 56: 153–166.

Lal, R., P. Henderlong, and M. Flowers. 1998. Forages and row cropping effects on soil organic carbon and nitrogen contents. In Lal, R. et al., Eds., *Management of C Sequestration in Soil.* CRC Press, Boca Roca, FL, pp. 365–379.

Larson, W.E. et al. 1972. Effects of increasing amounts of organic residues on continuous corn II. Organic carbon, nitrogen, phosphorus and sulfur. *Agron. J.* 64: 204–208.

Lehmann, J. and M. Rondon. 2006. Biochar soil management on highly weathered soils in the humid tropics. In Uphoff, N. et al., Eds., *Biological Approaches to Sustainable Soil Systems.* CRC Press, Boca Raton , FL, pp. 517–530.

Lehmann, J., J. Gaunt, and M. Rondon. 2006. Biochar sequestration in terrestrial ecosystems: Review. *Mitigation Adapt. Strat. Glob. Clim. Change* 11: 403–427.

Lehmann, J. et al. 2003. Nutrient availability and leaching in an archaeological anthrosol and a aerralsol of the Central Amazon Basin: Fertilizer, manure and charcoal amendments. *Plant Soil* 249: 343–357.

Lemus, R. et al. 2008. Effects of nitrogen fertilization on biomass yield and quality in large fields of established switchgrass in southern Iowa, USA. *Biom. Bioenergy* 32: 1187–1194.

Lemus, R. et al. 2002. Biomass yield and quality of 20 switchgrass populations in southern Iowa, USA. *Biom. Bioenergy* 23: 433–442.

Li, H. et al. 2007. Effects of 15 years of conservation tillage on soil structure and productivity of wheat cultivation in northern China. *Aust. J. Soil Res.* 45: 344–350.

Limon-Ortega, A. et al. 2006. Soil aggregate and microbial biomass in a permanent bed wheat–maize planting system after 12 years. *Field Crops Res.* 97: 302–309.

Limon-Ortega, A. et al. 2002. Soil attributes in a furrow-irrigated bed planting system in northwest Mexico. *Soil Tillage Res.* 63: 123–132.

Linden, D.R., C.E. Clapp, and R.H. Dowdy. 2000. Long-term corn grain and stover yields as a function of tillage and residue removed in east central Minnesota. *Soil Tillage Res.* 56: 167–174.

Lindstrom, M.J. 1986. Effects of residue harvesting on water runoff, soil erosion and nutrient loss. *Agric. Ecosyst. Environ.* 16: 103–112.

Linquist, B.A., V. Phengsouvanna, and P. Sengxue. 2007. Benefits of organic residues and chemical fertilizer to productivity of rain-fed lowland rice and to soil nutrient balances. *Nutr. Cycling Agroecosyst.* 79: 59–72.

Liu, W.D. et al. 2004. Response of corn grain yield to spatial and temporal variability in emergence *Crop Sci.* 44: 847–854.

Lobell, D.B. and G.P. Asner. 2002. Moisture effects on soil reflectance. *Soil Sci. Soc. Am. J.* 66: 722–727.

Low, A.J. 1972. The effect of cultivation on the structure and other physical characteristics of grassland and arable soils. *J. Soil Sci.* 23: 363–380.

Lynch, J. and E. Bragg. 1985. Microorganisms and soil aggregate stability. *Adv. Soil Sci.* 2: 133–171.

Ma, Z., C.W. Wood, and D.I. Bransby. 2000. Impacts of soil management on root characteristics of switchgrass. *Biom. Bioenergy* 18: 105–112.

Madakadze, I.C. et al. 1999. Cutting frequency and nitrogen fertilization effects on yield and nitrogen concentration of switchgrass in a short season area. *Crop Sci.* 39: 552–557.

Magleby, R. 2003. Soil Management and Conservation. USDA-ARS. http://www.ers.usda.gov/publications/arei/ah722/dbgen.htm (verified June 25, 2008).

Maiorano, A. et al. 2008. Effects of maize residues on the *Fusarium* spp. infection and deoxynivalenol (DON) contamination of wheat grain. *Crop Prot.* 27: 182–188.

Major, D.J., F.J. Larney, and C.W. Lilndwall. 1990. Spectral reflectance characteristics of wheat residues. In *Tenth Annual International Geoscience and Remote Sensing Symposium: Remote Sensing for the Nineties*. College Park, MD, May 1990, pp. 603–607.

Malhi, S.S. and H.R. Kutcher. 2007. Small grains stubble burning and tillage effects on soil organic C and N, and aggregation in northeastern Saskatchewan. *Soil Tillage Res.* 94: 353–361.

Malhi, S.S. and R. Lemke. 2007. Tillage, crop residue and N fertilizer effects on crop yield, nutrient uptake, soil quality and nitrous oxide gas emissions in a second 4-year rotation cycle. *Soil Tillage Res.* 96: 269–283.

Malhi, S.S. et al. 2006. Tillage, nitrogen and crop residue effects on crop yield, nutrient uptake, soil quality, and greenhouse gas emissions. *Soil Tillage Res.* 90: 171–183.

Malik, R.K. et al. 2000. Use of cover crops in short rotation hardwood plantations to control erosion. *Biom. Bioenergy* 18: 479–487.

Manlay, R.J. et al. 2002. Carbon, nitrogen and phosphorus allocation in agro-ecosystems of a West African savanna III. Plant and soil components under continuous cultivation. *Agric. Ecosyst. Environ.* 88: 249–269.

Mann, L., V. Tolbert, and J. Cushman. 2002. Potential environmental effects of corn (*Zea mays* L.) stover removal with emphasis on soil organic matter and erosion. *Agric. Ecosyst. Environ.* 89: 149–166.

Markvart, T. and L. Castañer, Eds. 2003. *Practical Handbook of Photovoltaics: Fundamentals and Applications*. Elsevier, New York.

Martens, D.A. 2000. Plant residue biochemistry regulates soil carbon cycling and carbon sequestration. *Soil Biol. Biochem.* 32: 361–369.

McCalla, T.M. 1943. Microbiological studies of the effect of straw used as a mulch. *Trans. Kan. Acad. Sci.* 46: 52–56.

McCool, D.K. et al. 2006. Residue characteristics for wind and water erosion control. 14th International Soil Conservation Organization Conference, Marrakech, Morocco, May 2006, p. 416. http://www.tucson.ars.ag.gov/isco/index_files/Page416.htm (verified Dec. 4, 2008).

McVay, K.A. et al. 2006. Management effects on soil physical properties in long-term tillage studies in Kansas. *Soil Sci. Soc. Am. J.* 70: 434–438.

Merrill, S.D. et al. 2006. Soil coverage by residue as affected by ten crop species under no-till in the northern Great Plains. *J. Soil Water Conserv.* 61: 7–13.

Mikha, M.M. and C.W. Rice. 2004. Tillage and manure effects on soil and aggregate-associated carbon and nitrogen. *Soil Sci. Soc. Am. J.* 68: 809–816.

Miller, R.M. and J.D. Jastrow. 1990. Hierarchy of root and mycorrhizal fungal interactions with soil aggregation. *Soil Biol. Biochem.* 22: 570–584.

Mitchell, R., J. Webb, and R. Harrison. 2001. Crop residues can affect N leaching over at least two winters. *Eur. J. Agron.* 15: 17–29.

Montgomery, D.R. 2007. Soil erosion and agricultural sustainability. *Proc. Natl. Acad. Sci. USA* 104: 13268–13272.

Monti, A., N. Di Virgilio, and G. Venturi. 2008. Mineral composition and ash content of six major energy crops. *Biom. Bioenergy* 32: 216–223.

Mubarak, A.R. et al. 2002. Decomposition and nutrient release of maize stover and groundnut haulm under tropical field conditions of Malaysia. *Commun. Soil Sci. Plant Anal.* 33: 609–622.

Munawar, A. et al. 1990. Tillage and cover crop management for soil water conservation. *Agron. J.* 82:773-777.

Nafziger, E.D., P.R. Carter, and E.E. Graham. 1991. Response of corn to uneven emergence. *Crop Sci.* 31: 811–815.

Nearing, M.A., F.F. Pruski, and M.R. O'Neal. 2004. Expected climate change impacts on soil erosion rate: Review. *J. Soil Water Conserv.* 59: 43–50.

Nelson, R.G. 2002. Resource assessment and removal analysis for corn stover and wheat straw in the Eastern and Midwestern United States: Rainfall- and wind-induced soil erosion methodology. *Biom. Bioenergy* 22: 349–363.

Nicholson, F.A. et al. 1997. Effects of repeated straw incorporation on crop fertiliser nitrogen requirements, soil mineral nitrogen and nitrate leaching losses. *Soil Use Mgt.* 13: 136–142.

Novak, M.D., W. Chen, and M.A. Hares. 2000. Simulating the radiation distribution within a barley–straw mulch. *Agric. For. Meteorol.* 102: 173–186.

Nuutinen, V. 1992. Earthworm community response to tillage and residue management on different soil types in southern Finland. *Soil Tillage Res.* 23: 221–239.

O'Halloran, I.P., M.H. Miller, and G. Arnold. 1986. Absorption of P by corn (*Zea mays* L.) as influenced by soil disturbance. *Can. J. Soil Sci.* 66: 287–302.

Oades, J.M. 1984. Soil organic matter and structural stability: Mechanisms and implications for management. *Plant Soil* 76: 319–337.

Parton, W.J. 1996. The CENTURY model. In Dowlson, D.S. et al., Eds., *Evaluation of Soil Organic Matter Models.* NATO ASI Series I, Vol. 38. Springer, Berlin, pp. 283–291.

Paul, E.A. 1991. Decomposition of organic matter. In Lederburg, J., Ed., *Encyclopedia of Microbiology*, Vol. 3. Academic Press, San Diego, CA, pp. 289–304.

Paustian, K., H.P. Collins, and E.A. Paul. 1997. Management controls on soil carbon. In Paul, E.A. et al., Eds., *Soil Organic Matter in Temperate Agroecosystems: Long-Term Experiments in North America.* CRC Press, Boca Raton, FL, pp. 15–49.

Paustian, K., W.J. Parton, and J. Persson. 1992. Modeling soil organic matter in organic-amended and nitrogen-fertilized long-term plots. *Soil Sci. Soc. Am. J.* 56: 476–488.

Perlack, R.D. et al. 2005. Biomass as feedstock for a bioenergy and bioproducts industry: Technical feasibility of a billion-ton annual supply. U.S. Department of Energy and U.S. Department of Agriculture. http://www.eere.energy.gov/biomass/pdfs/final_billionton_vision_report2.pdf (posted July 15, 2005; verified Sept. 4, 2008).

Pikul, J.L.J. et al. 2008. Change in surface soil carbon under rotated corn in eastern South Dakota. *Soil Sci. Soc. Am. J.* 72: 1738–1744.

Power, J.F., W.W. Wilhelm, and J.W. Doran. 1986. Crop residue effects on soil environment and dryland maize and soybean production. *Soil Tillage Res.* 8: 101–0111.

Power, J.F. et al. 1998. Residual effects of crop residues on grain production and selected soil properties. *Soil Sci. Soc. Am. J.* 62: 1393–1397.

Pratt, A.W. 1969. Heat transmission in low conductivity materials. In Tye, R.P., Ed., *Thermal Conductivity,* Vol. 1. Academic Press, New York, pp. 301–405.

Rachman, A. et al. 2004. Soil hydraulic properties influenced by stiff-stemmed grass hedge systems. *Soil Sci. Soc. Am. J.* 68: 1386–1393.

Rao, S.C., H.S. Mayeux, and B.K. Northup. 2005. Performance of forage soybean in the southern Great Plains. *Crop Sci.* 45: 1973–1977.

Rasmussen, P.E. et al. 1980. Crop residue influences on soil carbon and nitrogen in a wheat–fallow system. *Soil Sci. Soc. Am. J.* 44: 596–600.

Rathore, T.R., B.P. Ghildyal, and R.S. Sachan. 1982. Germination and emergence of soybean under crusted soil conditions II. Seed environment and varietal differences. *Plant Soil* 65: 73–77.

Reicosky, D.C. et al. 2002. Continuous corn with moldboard tillage: Residue and fertility effects on soil carbon. *J. Soil Water Conserv.* 57: 277–284.

Reynolds, J.H., C.L. Walker, and M.J. Kirchner. 2000. Nitrogen removal in switchgrass biomass under two harvest systems. *Biom. Bioenergy* 19: 281–286.

Rhoton, F.E., M.J. Shipitalo, and D.L. Lindbo. 2002. Runoff and soil loss from midwestern and southeastern U.S. silt loam soils as affected by tillage practice and soil organic matter content. *Soil Tillage Res.* 66: 1–11.

Rickman, R. et al. 2002. Tillage, crop rotation, and organic amendment effect on changes in soil organic matter. *Environ. Pollut.* 116: 405–411.

Roldan, A. et al. 2003. No-tillage, crop residue additions, and legume cover cropping effects on soil quality characteristics under maize in Patzcuaro watershed. *Soil Tillage Res.* 72: 65–73.

Ross, P.J., J. Williams, and R.L. McCown. 1985. Soil temperature and the energy balance of vegetative mulch in the semi-arid tropics II. Dynamic analysis of the total energy balance. *Aust. J. Soil Res.* 23: 515–532.

Russel, J.C. 1940. The effect of surface cover on soil moisture losses by evaporation. *Soil Sci. Soc. Am. J.* 4: 65–70.

Saffigna, P.G. et al. 1989. Influence of sorghum residues and tillage on soil organic matter and soil microbial biomass in an Australian vertisol. *Soil Biol. Biochem.* 21: 759–765.

Sainju, U.M. et al. 2006. Carbon sequestration in dryland soils and plant residue as influenced by tillage and crop rotation. *J. Environ. Qual.* 35: 1341–1347.

Salinas-Garcia, J.R. et al. 2001. Residue removal and tillage interaction effects on soil properties under rain-fed corn production in Central Mexico. *Soil Tillage Res.* 59: 67–79.

Sarr, P.S. et al. 2008. Effect of pearl millet–cowpea cropping systems on nitrogen recovery, nitrogen use efficiency and biological fixation using the ^{15}N tracer technique. *Soil Sci. Plant Nutr.* 54: 142–147.

Sauer, T.J. et al. 1998. Surface energy balance of a corn residue-covered field. *Agric. For. Meteorol.* 89: 155–168.

Shafii, B. and W. Price. 2001. Estimation of cardinal temperatures in germination data analysis. *J. Agric. Biol. Environ. Stat.* 6: 356–366.

Sharratt, B.S. 2002. Corn stubble height and residue placement in the northern U.S. Corn Belt I. Soil physical environment during winter. *Soil Tillage Res.* 64: 243–252.

Sharratt, B.S. and G.S. Campbell. 1994. Radiation balance of a soil-straw surface modified by straw color. *Agron. J.* 86: 200–203.

Sharratt, B.S., G.R. Benoit, and W.B. Voorhees. 1998. Winter soil microclimate altered by corn residue management in the northern Corn Belt of the USA. *Soil Tillage Res.* 49: 243–248.

Shen, Y.J. and C.B. Tanner. 1990. Radiative and conductive transport of heat through flail-chopped corn residue. *Soil Sci. Soc. Am. J.* 54: 653–658.

Singh, B. and S.S. Malhi. 2006. Response of soil physical properties to tillage and residue management on two soils in a cool temperate environment. *Soil Tillage Res.* 85: 143–153.

Singh, B. et al. 1994. Residue and tillage management effect on soil properties of a typic cryoboroll under continuous barley. *Soil Tillage Res.* 32: 117–133.

Six, J., E.T. Elliott, and K. Paustian. 1999. Aggregate and soil organic matter dynamics under conventional and no-tillage systems. *Soil Sci. Soc. Am. J.* 63: 1350–1358.

Six, J., E.T. Elliott, and K. Paustian. 2000a. Soil macroaggregate turnover and microaggregate formation: Mechanism for C sequestration under no-tillage management. *Soil Biol. Biochem.* 32: 2099–2103.

Six, J. et al. 1998. Aggregation and soil organic matter accumulation in cultivated and native grassland soils. *Soil Sci. Soc. Am. J.* 62: 1367–1377.

Six, J. et al. 2000b. Soil structure and organic matter I. Distribution of aggregate-size classes and aggregate-associated carbon. *Soil Sci. Soc. Am. J.* 64: 681–689.

Skidmore, E.L. 1988. Wind erosion. In Lal, R., Ed., *Soil Erosion Research Methods*. Soil Water Conservation Society, Ankeny, IA, pp. 203–233.

Smika, D.E. and P.W. Unger. 1986. Effect of surface residues on soil water storage. *Adv. Soil Sci.* 5: 111–138.

Smiley, R.W., H.P. Collins, and P.E. Rasmussen. 1996. Diseases of wheat in long-term agronomic experiments at Pendleton, Oregon. *Plant Dis.* 80: 813–820.

Soane, B.D. 1990. The role of organic matter in soil compactibility: Review of some practical aspects. *Soil Tillage Res.* 16: 179–201.

Spedding, T.A. et al. 2004. Soil microbial dynamics in maize-growing soil under different tillage and residue management systems. *Soil Biol. Biochem.* 36: 499–512.

Stanhill, G., G.J. Hofstede, and J.D. Kalma. 1966. Radiation balance of natural agricultural vegetation. *Q. J. Roy. Meteorol. Soc.* 92.

Steiner, J. 1994. Crop residue effects on water conservation. In Unger, P.W., Ed., *Managing Agricultural Residues.* CRC Press, Boca Raton, FL, pp. 41–76.

Steiner, J.L. et al. 2000. Biomass and residue cover relationships of fresh and decomposing small grain residue. *Soil Sci. Soc. Am. J.* 64: 2109–2114.

Stevenson, F.J. 1994. *Humic Chemistry: Genesis, Composition, Reactions*, 2nd ed. John Wiley & Sons, New York.

Stocking, M.A. 1988. Assessing vegetative cover and management effects. In Lal, R., Ed., *Soil Erosion Research Methods.* Soil Water Conservation Society, Ankeny, IA, pp. 211–234.

Swan, J.B. et al. 1994. Surface residue and in-row treatment on long-term no-tillage continuous corn. *Agron. J.* 86: 711–718.

Tanner, C.B. and Y. Shen. 1990. Solar-radiation transmittance of flail-chopped corn residue layers. *Soil Sci. Soc. Am. J.* 54: 650–652.

Tian, G., B.T. Kang, and L. Brussand. 1992. Biological effects of plant residues with contrasting chemical composition under humid tropical conditions: Decomposition and nutrient release. *Soil Biol. Biochem.* 24: 1051–1060.

Tirol-Padre, A., K. Tsuchiya, K. Inubushi, and J.K. Ladha. 2005. Enhancing soil quality through residue management in a rice–wheat system in Fukuoka, Japan. *Soil Sci. Plant Nutr.* 51: 849–860.

Tisdall, J.M. 1994. Possible role of soil microorganisms in aggregation in soil. *Plant Soil* 159: 115–121.

Tisdall, J.M. 1996. Formation of soil aggregates and accumulation of soil organic matter. In Carter, M.R. and B.A. Stewart, Eds., *Structure and Organic Matter Storage in Agricultural Soils.* CRC Press, Boca Raton, FL, pp. 57–96.

Tisdall, J.M. and J.M. Oades. 1982. Organic matter and water-stable aggregates in soils. *J. Soil Sci.* 33: 141–163.

Tomer, M.D., D.W. Meek, and L.A. Kramer. 2005. Agricultural practices influence flow regimes of headwater streams in western Iowa. *J. Environ. Qual.* 34: 1547–1558.

Tomer, M.D. et al. 2006. Surface soil properties and water contents across two watersheds with contrasting tillage histories. *Soil Sci. Soc. Am. J.* 70: 620–630.

Uhlenbrook, S. 2007. Biofuel and water cycle dynamics: What are the related challenges for hydrological processes research? *Hydrol. Proc.* 21: 3647–3650.

Updegraffa, K., P. Gowdaa, and D.J. Mullaa. 2004. Watershed-scale modeling of the water quality effects of cropland conversion to short-rotation woody crops. *Renew. Agricult. Food Syst.* 19: 118–127.

USDA NRCS. 1997. Tolerable Erosion Levels. USDA-NRCS. http://www.mn.nrcs.usda.gov/technical/nri/findings/erosion_t.htm (verified Dec. 1, 2008).

van Bavel, C.H.M. and D.I. Hillel. 1976. Calculating potential and actual evaporation from a bare soil surface by simulation of concurrent flow of water and heat. *Agric. Meteorol.* 17: 453–476.

Van Donk, S.J. and E.L. Skidmore. 2003. Measurement and simulation of wind erosion, roughness degradation and residue decomposition on an agricultural field. *Earth Surf. Proc. Landforms* 28: 1243–1258.

Van Veen, J.A. and P.J. Kuikman. 1990. Soil structural aspects of decomposition of organic matter by microorganisms. *Biogeochemistry* 11: 213–223.

Van Veen, J.A. and E.A. Paul. 1981. Organic carbon dynamics in grassland soils 1: Background information and computer simulation. *Can. J. Soil Sci.* 61: 185–201.

Vanotti, M.B., L.G. Bundy, and A.E. Peterson. 1997. Nitrogen fertilizer and legume–cereal rotation effects on soil productivity and organic matter dynamics in Wisconsin. In Paul, E.A. et al., Eds., *Soil Organic Matter in Temperate Agroecosystems.* CRC Press, Boca Raton, FL, pp. 105–119.

Varvel, G.E., and W.W. Wilhelm. 2008. Soil carbon levels in irrigated Western Corn Belt rotations. *Agron. J.* 100: 1180–1184.

Velthof, G.L., P.J. Kuikman, and O. Oenema. 2002. Nitrous oxide emission from soils amended with crop residues. *Nutr. Cycl. Agroecosys.* 62: 249–261.

Vitosh, M.L., R.E. Lucas, and G.H. Silva. 1997. Long-term effects of fertilizer and manure on corn yield, soil carbon, and other soil chemical properties in Michigan. In Paul, E.A. et al., Eds., *Soil Organic Matter in Temperate Egroecosystems.* CRC Press, Boca Raton, FL, pp. 129–139.

Vogel, K.P. et al. 2002. Switchgrass biomass production in the midwest USA: Harvest and nitrogen management. *Agron. J.* 94: 413–420.

Wagner-Riddle, C. et al. 2008. Linking nitrous oxide flux during spring thaw to nitrate denitrification in the soil profile. *Soil Sci. Soc. Am. J.* 72: 908–916.

Wang, W.J. et al. 2004. Decomposition dynamics of plant materials in relation to nitrogen availability and biochemistry determined by NMR and wet-chemical analysis. *Soil Biol. Biochem.* 36: 2045–2058.

Wells, R.R. et al. 2003. Infiltration and surface geometry features of a swelling soil following successive simulated rainstorms. *Soil Sci. Soc. Am. J.* 67: 1344–1351.

Wieder, R.K. and G.E. Lang. 1982. A critique of analytical methods used in examining decomposition data obtained from litter bags. *Ecology* 63: 1636–1642.

Wilhelm, W.W., J.W. Doran, and J.F. Power. 1986. Corn and soybean yield response to crop management under no-tillage production systems. *Agron. J.* 7: 184–189.

Wilhelm, W.W., J.M.F. Johnson, D.L. Karlen, and D.T. Lightle. 2007. Corn stover to sustain soil organic carbon further constrains biomass supply. *Agron. J.* 99: 1665–1667.

Wilhelm, W.W. et al. 2004. Crop and soil productivity response to corn residue removal: A literature review. *Agron. J.* 96: 1–17.

Wilson, G.V., K.C. McGregor, and D. Boykin. 2008. Residue impacts on runoff and soil erosion for different corn plant populations. *Soil Tillage Res.* 99: 300–307.

Wilts, A.R., D.C. Reicosky, R.R. Allmaras, and C.E. Clapp. 2004. Long-term corn residue effects: Harvest alternatives, soil carbon turnover, and root-derived carbon. *Soil Sci. Soc. Am. J.* 68: 1342–1351.

Wuest, S.B. et al. 2005. Organic matter addition, N, and residue burning effects on infiltration, biological, and physical properties of an intensively tilled silt-loam soil. *Soil Tillage Res.* 84: 154–167.

Yaman, S. 2004. Pyrolysis of biomass to produce fuels and chemical feedstocks. *Energy Conv. Mgt.* 45: 651–671.

Yang, H.S., D.J. Kim, and H.J. Kim. 2003. Rice straw–wood particle composite for sound absorbing wooden construction materials. *Bioresour. Technol.* 86: 117–121.

Ying, J. et al. 1998. Comparison of high-yield rice in tropical and subtropical environments II. Nitrogen accumulation and utilization efficiency. *Field Crops Res.* 57: 85–93.

Yu, F., R. Ruan, and P. Steele. 2008. Consecutive reaction model for the pyrolysis of corn cob. *Trans. ASABE* 51: 1023–1028.

Zan, C.S. et al. 2001. Carbon sequestration in perennial bioenergy, annual corn and uncultivated systems in southern Quebec. *Agric. Ecosyst. Environ.* 86: 135–144.

Zemenchik, R.A. et al. 2000. Corn production with Kura clover as a living mulch. *Agron. J.* 92: 698–705.

Zhang, G.S. et al. 2007. Relationship between soil structure and runoff/soil loss after 24 years of conservation tillage. *Soil Tillage Res.* 92: 122–128.

2 Soil Quality Impacts of Residue Removal for Biofuel Feedstock

Richard M. Cruse, M.J. Cruse, and D.C. Reicosky

CONTENTS

INTRODUCTION

With increased tensions among political powers in countries controlling fossil fuel supplies, the debate about the reliability of these sources to meet future global demand continues to escalate. Taking into consideration the ever-present implications related to global warming, the world's energy paradigm is becoming increasingly heterogenic. Included in the mixture of energy sources to be utilized are renewable sources such as plant residues. Crop and non-crop plant residues have been repeatedly identified as plentiful renewable resources for cellulosic biofuel production (Perlack et al., 2005) and have been targeted as significant contributors to the future energy portfolios of the United States and other countries (Wiesenthal et al., 2006).

Plant residues are frequently considered free products of photosynthesis and carbon capture whose ecological function and economic value are only marginally important. Therefore, to many, plant residues appear to be prime candidates for conversion into economically valuable products that may help reduce our dependency on foreign oil and invigorate rural economies. Under this assumption, utilizing plant residue feedstocks for renewable liquid fuel production offers multiple opportunities with few apparent drawbacks. Unfortunately, these opportunities are harnessed with at least one often unrecognized major conflict: The plant carbon used in liquid fuel production is the same plant carbon that is a critical foundation of soil quality, and soil is the fundamental foundation of our economy and our existence.

Soil quality strongly depends on soil carbon (Lal, 2005). Degradation of soil resources and the associated potential productivity loss would more than negate gains attributed to liquid fuel production from plant residue feedstocks. Multiple studies have tied excessive above-ground residue removal to soil quality degradation and loss in soil and cropping system performance (Mann, et al., 2002; Wilhelm et al., 2004; Wilhelm et al., 2007; Johnson et al., 2006a; Lal, 2004 and 2005; Lemus and Lal, 2005; Reicosky and Wilts, 2004). The dilemma soil scientists face is identifying appropriate levels of crop residue removal that will support national goals presented in the Energy Independence and Security Act of 2007 and concurrently maintain soil quality. Because healthy soil is required for food, feed, fiber, and fuel production, the importance of soil health is arguably comparable to that of renewable energy production for meeting national security goals. This chapter's purpose is to provide an overview of the potential soil quality and environmental implications associated with crop biomass removal for bioenergy production from our agricultural landscapes. It will address soil C and ecosystem function as they relate to environmental services and the potential impact of reducing C soil additions via biomass harvest. Throughout the following discussion, the terms *soil C* and *SOM content* are used interchangeably. See other reviews of the role of C sequestration in conservation agriculture presented by Robert (2001), Uri (1999), Tebrugge and Guring (1999), Lal et al. (1998), and Lal (2000).

SOIL QUALITY AND ECOSYSTEM FUNCTION

The Soil Society of America defines soil quality as "the capacity of a soil to function within ecosystem boundaries to sustain biological productivity, maintain environmental quality, and promote plant and animal health" (https://www.soils.org/sssa-gloss/index.php). Although the universal applicability of the soil quality concept has been debated in the literature (Sojka and Upchurch, 1999), the impacts of processes within the soil and production from the soil are well understood (Doran et al., 1994; Doran and Jones, 1996). As a growing global population places higher demands on soils, maintaining or improving soil quality is increasingly critical.

Soils have historically been recognized as production media for food, feed, fiber, and fuel; as ecosystem components that effectively decompose wastes and recycle nutrients; as water storage media; and as filtering mechanisms for water moving through the profile. As atmospheric greenhouse gas concentrations have increased, soils have been recognized as resources for carbon sequestration and climate change mitigation (Lal et al., 1998). Now, soils are being asked to produce liquid fuel feedstocks (ethanol and biodiesel) from grain and oil seed crops and will soon be asked to produce plant biomass for cellulosic ethanol and other types of liquid fuel conversions. Of the listed demands placed on soils, production and removal of biomass as liquid fuel feedstocks offer the greatest potential threat to soil resources, as will be described. Continued degradation of this finite resource that provides so many essential products for an expanding global population is not only shortsighted—it is very dangerous.

SOIL ORGANIC MATTER

Multiple soil components interact, influencing soil function; but no component is more important than soil carbon (C)—a term often used interchangeably with soil organic matter (SOM). Soil carbon is dynamic and responds to changes in tillage, C input through root and root exudates, soil incorporation of above-ground plant materials (Johnson et al., 2006), and additions of organic materials such as manure (Lal, 2004). Carbon bonds within soil organic matter are the main determinants of biological activity because they are the primary energy sources for living soil organisms. The amount, diversity, and activity of soil fauna and flora are directly related to SOM quantity and quality. Organic matter and the biological activity that it generates exert a major influence on soil physical properties, chemical properties, and nutrient cycling.

BIOMASS HARVEST AND SOIL ORGANIC MATTER

Focal points associated with bioenergy cropping systems, SOM, and the C cycle involve (1) CO_2 capture in photosynthesis and CO_2 release associated with fossil fuel compared with biofuel use and (2) the impact of biomass harvest on SOM and the resulting changes in soil productivity and ecosystem function (Johnson et al., 2006a; Bransby et al., 1998; Lal et al., 1998; Lal, 2004 and 2005; Lemus and Lal, 2005; Wilhelm et al., 2004). Biofuels are renewable and can favorably affect atmospheric CO_2 concentrations compared to concentrations resulting from fossil fuel combustion (Farrell et al., 2006). However, soils have already lost much of their original C pools through intensive conventional agricultural (Lal, 2002), and concerns exist that poorly managed biomass feedstock production systems will further degrade our finite soil resource.

Improvement in soil quality with crop biomass removal for bioenergy feedstocks is a very complex technical challenge. It depends on identifying crop species (1) capable of high productivity (carbon capture) with low fertilization requirements; (2) amenable to available planting and harvest management techniques; (3) with improved crop rooting characteristics; and (4) allowing landscape preservation with minimal environmental impact. Fortunately, with the adoption of selected high biomass producing species, 60% to 70% of the current depleted SOM pool may be resequestered (Lal, 2002) and favorable atmospheric CO_2 balance may result (Farrell et al., 2006).

Dedicated perennial biomass systems may reduce net atmospheric CO_2 releases and increase SOM content on most farmed soils. However, if we continue utilizing our current corn–soybean row cropping practices and simply remove above-ground biomass as bioenergy feedstock, the potential to reduce SOM content and degrade soil quality increases. Multiple studies reinforce this contention and illustrate the negative impact on SOM of above-ground corn biomass removal for silage (Reicosky et al., 2002; Karlen et al., 1994; Robinson et al., 1996; Vitosh et al., 1997). Unfortunately, long-term field studies have not conclusively documented SOM content impacts of partial above-round biomass harvest for crops such as corn (*Zea mays* L.). Wilhelm et al. (2007) suggest that corn residue quantities required in the field to maintain SOM levels will average substantially greater than amounts required to keep soil erosion at or below defined tolerable levels.

Biomass removal impacts on SOM loss may be at least partially offset by manure additions; in studies not involving biomass removal, substantial manure applications maintained or increased SOM within different cropping systems (Kanchikerimath and Singh, 2001; Anderson et al., 1990; and Sommerfeldt et al., 1988). The assumption that this will happen with biomass harvest is somewhat conjecture as solid long-term field data cannot yet quantify the effect of manure addition on SOM in conjunction with biomass removal. Additionally, commercial farms couple silage removal with live-stock systems; silage is fed to livestock and manure is returned to fields from which the silage was removed. Many harvest and feedstock conversion systems for bioenergy feedstocks will unlikely return nutrient rich by-products such as manure to the soil or the conversion facility may not be close enough to the source of the ash and biochar material to economically justify hauling it to the field from which it came.

NUTRIENT CYCLING AND CATION EXCHANGE

Ion adsorption or exchange is one of the most important soil functions related to nutrient cycling and an important means by which plant nutrients are retained in crop rooting zones. SOM and its mineralization play a significant role in this soil function. Crovetto (1996) showed that the contribution of SOM to cation exchange capacity exceeded that of the kaolinite clay mineral in the surface 5 cm of an alfisol (haploxeralfs) in the Chequén region of Chile. Robert (1996 and 2001) showed a strong linear relationship between organic C and cation exchange complex (CEC) of his experimental soil. The CEC increased fourfold with an organic C increase from 1% to 4%. Quantitative evaluation of biomass removal impacts on CEC is not yet available, but the literature indicates a direct relationship between SOM content and CEC. Coupling this with literature suggesting that biomass removal reduces SOM content (Wilhelm et al., 2007) leads us to conclude biomass harvest will lower soil CEC for at least some, if not most soils.

NUTRIENT BALANCE AND CONTENT

Crop residues contain significant quantities of several nutrients. Most crops concen-trate nutrients in their seeds, but significant quantities of nutrients remain in most crop residues and upon decomposition of the residue in soil are slowly released for plant uptake, incorporation into microbial biomass, adsorption into the cation exchange complex, or simply remain as part of the more stable soil organic matter fraction. Removal of these residues and nutrients contained therein and losses associated with

TABLE 2.1
Cost Estimates for Nutrients Removed from Crop Residues

	Concentration (g kg⁻¹)				
Crop	Nitrogen	Phosphorus	Potassium	Nutrient Replacement Cost $/Tonne	Source
Corn stover	9.83	1.00	15.04	21.12	NRCS, http://www.nrcs.usda.gov/TECHNICAL/ECS/nutrient/tbb1.html
Switchgrass	9.44	0.90	19.08	23.16	Fixen, 2007
Wheat straw	6.11	0.64	11.74	14.72	NRCS, http://www.nrcs.usda.gov/TECHNICAL/ECS/nutrient/tbb1.html
Oat straw	7.08	0.85	23.89	24.10	http://www.nrcs.usda.gov/TECHNICAL/ECS/nutrient/tbb1.html

Note: Prices of N, P, and K based on estimated costs of crop production in Iowa (Duffy, 2008) are $0.46, $0.50, and $0.27 per pound of N, P_2O_5, and K_2O, respectively.

accelerated erosion from less surface cover (Larson et al., 1978) reduce soil nutrient availability (Barnhart et al., 1978) and increase supplemental fertilizer requirements for subsequent crops. Table 2.1 lists macronutrient concentrations of selected residues considered for biofuel feedstocks and estimated costs of replacing these nutrients per ton of residue removed. The added fertilizer cost to the producer is substantial because such costs have increased dramatically (Fixen, 2007).

SOIL STRUCTURE

Soil aggregates, the primary soil structure components, are composed of mineral and organic matter plus gases and water within the aggregate voids. Aggregates are generally orders of magnitude larger than the primary particles from which they are made. Through natural aggregation of primary particles into these structural units, various soil physical, biological and chemical properties and processes are changed and generally improved. Well-structured soils exhibit better water infiltration (Franzluebbers, 2002), better gas exchange (Edwards et al., 1993), reduced root growth resistance (Lampurlanés and Cantero-Martínez, 2003), and lower soil erosion rates (Burwell and Larson, 1969).

Organic matter additions to soil have long been known to improve soil structure, i.e., lead to increased soil aggregation. Various studies illustrate the correlation between soil structure and organic matter additions (Tisdall and Oades, 1982; Six et al., 2000; Mikha and Rice, 2004). Soil organic matter (1) improves internal and external binding of soil aggregates; (2) increases soil elasticity and rebounding capabilities following load applications; and (3) changes soil internal friction (Soane, 1990). If SOM is removed through oxidation, soil structural stability can be destroyed or substantially reduced (Chesire et al., 1983; Williams et al., 1966). The role of SOM in affecting soil physical condition is well-documented and largely unquestioned (Doran et al., 1994; Doran and Jones, 1996). Practices that remove plant material from the potential pool of SOM will likely negatively affect soil structure through effects on soil aggregation. If above-ground crop residue removal is to be practiced, sufficient material must remain in the field to foster aggregation processes (Johnson et al., 2006a; Wilhelm et al., 2007) or management practices such as no-till must be used to maintain the existing SOM resources (Larney et al., 1997; Reicosky and Archer, 2007).

Water Infiltration and Erosion

Soil erosion involves three processes: soil particle detachment from the primary soil body, transport of most of these detached particles in runoff water, and deposition at distance from the point of detachment. To effectively control soil erosion, we must interrupt either or both of the first two processes. Impacting raindrops represent the major energy source for soil detachment associated with water erosion. Crop residues effectively intercept raindrop energy, greatly reducing detachment and soil erosion during rainstorms (Cruse et al., 2001; Laflen and Colvin, 1981). Crop residues can also increase infiltration and therefore minimize erosion losses (Mannering and Meyer, 1963). As surface residues are removed and surface cover is reduced, such as may occur with biomass harvest, soil susceptibility to erosion is greatly increased (see Figure 2.1). With biomass removal, soil aggregate stability, and soil resistance to detachment forces become increasingly important for controlling the erosion process.

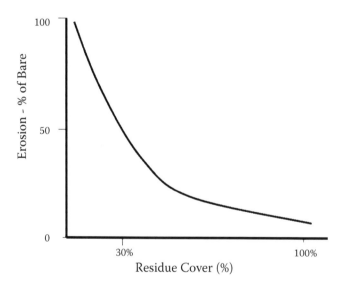

FIGURE 2.1 Effect of soil surface residue cover on soil erosion loss. Erosion is expressed as percent of erosion occurring with zero residue cover. (*Source:* Adapted from Laflen, J.M. and T.S. Colvin. 1981. *Trans. ASAE* 24: 605–609.)

SOM reduces soil erodibility by stabilizing surface aggregates (Bissonnias, 1996; Barthès and Roose, 2002). When a raindrop impacts an unprotected wet, weakened soil surface, particles are detached from the primary soil aggregate or clod (Cruse and Larson, 1977; Al-Durrah and Bradford, 1981) and transported in the splashing water droplets. The splashed soil particles pack into and plug larger open pores, decreasing surface layer porosity, reducing effective pore sizes and greatly reducing surface layer infiltration capacity. This reduces infiltration, particularly during heavy rain events, increases runoff rates and sediment transport potential, and leads to elevated soil erosion (Kemper and Miller, 1974). Soil organic matter stabilizing aggregates is especially important at very high soil water matric potentials (Francis and Cruse, 1983), when rainfall exceeds infiltration and runoff is likely.

The erosion process alone reduces soil quality through removal of the surface soil layer, typically containing soil with the highest SOM content, and degradation of the surface structural condition from open porous to a more sealed condition. Crop residues upon microbial decomposition help stabilize soil aggregates, reducing erosion, and if left on the soil surface prior to decomposition reduce erosion by both wind and water (Unger, 1994). Relative to erosion by water, residues absorb the impacts of raindrops and slow water movement over the soil surface, giving water more time to infiltrate (Jones et al., 1994). Relative to wind erosion, crop residues reduce velocity of air movement at the soil surface and can play a major role in reducing soil loss rates (Bilbro and Fryrear, 1994).

Thus, crop residue or biomass removal can impact infiltration and erosion through two primary mechanisms: (1) loss of soil surface protection from forces of wind and impacting rain and (2) alterations in aggregate stability and soil structure. Amounts of residue necessary to accommodate sustainability of each mechanism will likely differ (Wilhelm, 2007). Further, the amount of material required for each will differ across the landscape (Wilhelm et al., 2007). Spatial identifications of how much residue must remain at a given location and how much can be removed are necessary to optimize residue harvest quantities and soil quality maintenance. The spatial complexity involving residue yields and residue needs for surface soil quality maintenance and erosion control make this important management tool challenging to implement.

WATER AVAILABILITY

Water availability for crop growth is controlled by the quantity of water added to the soil profile via infiltration or irrigation and the quantity lost through drainage and evaporation. SOM, through its effect on soil structure and undecomposed surface crop residues, affects each of these processes as discussed elsewhere in this chapter. Surface residues and stable surface structure improve water infiltration, increasing the potential amount available for crop growth. SOM has long been considered effective for improving soil capacity to hold water against drainage.

Hudson (1994) showed that for each 1% increase in SOM, the soil available water-holding capacity increased volumetrically by 3.7%. In all texture groups of his study, as SOM content increased from 0.5% to 3%, available water capacity more than doubled. More recently, Olness and Archer (2005) estimated increases of 2% to 5% in available water capacity for each 1% increase in SOM. Practices that reduce SOM may cause more than trivial reductions in available water storage and may exert significant impacts on soil productivity, particularly in dry land situations.

Soil Water Evaporation

Similar to its effect on infiltration, organic materials originating from crop biomass affect evaporation internally through changes in soil structure and externally by potentially affecting gas exchange processes on the soil surface. Since soil water evaporation is primarily a soil surface process, movement of infiltrating water away from the soil surface to deeper levels during rainfall should reduce evaporation losses (Gardner and Gardner, 1969; Bresler and Kemper, 1972). In fact, the impact of SOM through its soil structure activity and subsequent evaporative losses of soil water may be very important (Lemon, 1956; Willis and Bond, 1971; Holmes et al., 1960).

Well-structured soils generally exhibit high saturated conductivity values compared to those of poorly structured soils of comparable texture. Higher saturated conductivity promotes deeper water movement more quickly under heavy rainfall; this favors water storage rather than evaporation when rainfall stops. Additionally, evaporation losses during stage 2 evaporation (Idso et al., 1974) will likely be reduced by well-structured conditions (Hartman et al., 1981), again favoring water storage.

Carbon Sequestration

World soils that constitute an important pool of active C play a major role in the global C cycle and contribute to changes in the atmospheric concentrations of greenhouse gases (Lal et al., 1998). Soil contains two to three times more C than the atmosphere (Council for Agriculture and Technology, 2004), even though intensive agriculture has caused a 30% to 50% loss of soil carbon in the past 120 years. If properly managed U.S. crop lands could sequester more carbon annually than is released by all U.S. agricultural activities.

Soil carbon sequestration in agricultural soils is directly related to SOM content, which again is mediated by management practices. Organic matter additions, typically through crop residues or animal manures, are positively correlated to SOM content (Lal, 2001). Practices such as plow and inversion tillage (Reicosky and Lindstrom, 1993; Reicosky, 1997; Ellert and Janzen, 1999; Stockfisch et al., 1999), artificial drainage (Baker et al., 2006), and high nitrogen fertilizer additions (Green et al., 1995; Robinson et al., 1996; Russell et al., 2005; Kahn et al., 2007) may influence biomass production and oxidation rates of organic materials added to soil, and thus affect total carbon sequestration. Little disagreement exists regarding crop residue importance in soil carbon sequestration. A current estimation is that 20% of the carbon produced by residue remains in the soil after two years and that 25% of soil-sequestered carbon in an agricultural system is derived from crop residue carbon (Wilhelm et al., 2004).

Cropping Effects on Soil Carbon

Clearly, attempting to meet projected feedstock requirements for biofuel production (Perlack et al., 2005) through residue removal from our current row cropping systems will negatively affect soil quality. Can we produce feedstocks for biofuels and maintain or improve soil quality, or is it impossible to avoid degradation of this critical resource?

The answer is based a number of considerations. We will briefly focus on two of critical importance: (1) direct management of the soil, e.g., tillage and applications of amendments may be altered for biomass harvest scenarios, and these management changes may impact soil quality independent of residue removal (Hooker et al., 2005); and (2) plant roots are the main sources of soil carbon in agricultural soils, and masses of organic materials contributed by roots vary considerably between existing and potential dedicated biomass crops (Allmaras et al., 2004; Johnson et al., 2006b; Balesdent and Balabane, 1992, 1996).

SOIL MANAGEMENT

A substantial amount of soil carbon comes from root systems, and for a variety of plant species, roots are the greatest soil carbon sources (Johnson, et al., 2006; Allmaras et al., 2004). What is gained by managing the rooting environment for maximum organic matter conservation (limited or no tillage) may balance or partially balance carbon lost through residue harvesting (Moebius-Clune, 2008). For example, in the northern Corn Belt, slow early growth and lower no-till corn yields have been attributed to cold or wet soils in the spring caused, at least in part, by crop residue cover (Kaspar et al., 1990). Removal of a portion of the crop residue as biofuel feedstock could reduce this no-till production concern, increase farmer adoption of no-till practices, and thereby reduce SOM losses attributed to more intensive tillage systems (Reicosky et al., 2002).

Other management practices affecting SOM content may also be altered. As noted earlier, removal of crop residues will likely result in increased fertilizer applications. Nitrogen fertilizers have been implicated in accelerating SOM loss (Khan et al., 2007), but are also credited for higher SOM levels when applied to soils on which nitrogen is limiting for optimum crop growth (Robinson et al., 1996). Judicious use of nitrogen seems the best option for maximizing production and minimizing potential SOM loss attributed to excess N fertilizer applications.

Biochar, a fine-grained char-like material high in organic carbon, largely resistant to decomposition, and a co-product of pyrolysis, may improve soil quality when applied as a soil amendment at appropriate rates (Laird, 2008). While research about biochar is arguably in its infancy, the material reportedly exhibits favorable impacts on the ability of soil to retain nutrients and absorb potential chemical pollutants, offers a means to sequester carbon derived from plant materials, and may increase soil water holding capacity (Laird, 2008).

The chemical properties of char are highly variable, and thus the impact on soil quality is expected to be variable as well. The properties of char depend strongly on the properties of the feedstock material and the operational temperature and pressure of the pyrolysis process used to form it. There is a need for explicit characterization of chemical and physical properties of this co-product based on certain benchmark materials (Goldberg, 1985; Schmidt and Noack, 2000; Lehman et al., 2003a and 2003b).

While mounting evidence indicates that biochar can favorably impact soil, application of such materials is not a guarantee of improved soil quality. Recent work by Wardle et al., (2008) describes the impact of fire-derived charcoal that causes loss of forest humus. Their results suggest that charcoal can promote rapid loss of forest humus and below-ground carbon during the first decade after its formation. The effect of charcoal on native soil carbon must be thoroughly investigated to better clarify the potential of charcoal carbon as an ecosystem sink and an agent for carbon sequestration. Additionally, based on anticipated soil quality impacts (Laird, 2008), biochar seems most likely to produce a positive impact on degraded soils; soils with favorable SOM contents and relatively high quality may not require substantial quality alterations through char applications since they are already in good condition.

CROP MANAGEMENT

If the cellulosic biofuel industry develops as anticipated, it may utilize a variety of cellulose sources for biofuel feedstock. Plant or crop species could range from existing crop residues such as those from corn stover and cobs to fast growing trees. The selection of crop species for feedstock will impact soil quality directly for a number of reasons, including fertility needs (especially N), tillage requirements, and root growth and carbon release dynamics. As already discussed, soil quality is sensitive to tillage and nutrient applications—especially nitrogen fertilizers. Requirements for these management activities and inputs differ among potential feedstock species. Those managed with less intensive tillage and lower N applications will likely undergo less SOM oxidation than cropping systems that require more. In fact, cropping systems such as dedicated perennial grasses will likely increase SOM even if biomass is harvested (Liebig, et al., 2008), especially if produced on land that was previously row cropped with significant tillage (Zan et al., 2001; Hooker et al., 2005).

Of equal importance to management variables is the role that crop root systems play in contributing carbon to soil (Allmaras et al., 2004; Johnson, et al., 2006b; Balesdaent and Balabane, 1992 and 1996). Root systems of crops differ considerably in the amounts and forms of carbon they release and contribute to a soil profile. Johnson et al. (2006b) reviewed the literature and illustrated that root biomasses and exudates contribute more organic carbon to soil than above-ground biomass. The proportions of SOM from above- and below-ground sources vary among species. Wilhelm et al. (2004) suggests as much as 75% of carbon found in soil comes from root sources. This should send the research community a strong signal that better understanding of below ground processes will be required to ultimately understand cropping effects on SOM content and soil quality, with or without residue removal.

While most soil carbon originates from roots as root biomass and rhizodeposition (Allmaras et al., 2004; Johnson et al., 2006b; Balesdent and Balabane, 1992 and 1996), caution is advised in making conclusions about its impacts on soil ecosystem function. Water infiltration and runoff, resistance to erosion, and soil aeration are controlled (or strongly affected) by the surface layer, and the deeper subsurface layers are affected by root biomass and rhizodeposition. Carbon sequestration seems influenced mostly by root-derived carbon; surface processes are highly affected by above-ground biomass residing on or mixed with shallow surface layers. Crop species fix different amounts of carbon and also distribute it differently above- and below ground (Figure 2.2; Tufekcioglu et al., 2003). Differences in leaf litter residing on the soil surface may play a little recognized but important role in soil ecosystem function.

Soil quality and effective ecosystem function can be viewed as a continuum with surface phenomena most strongly influenced by above-ground biomass removal that affects protection of soil from impacting rainfall and surface layer microbial residue decomposition processes. Research has not yet determined the quantity of above ground biomass necessary to maintain or improve shallow layer ecosystem functions for a range of soils and climatic conditions, but cited research indicates the amount is greater than zero. This determination should be a research priority of soil and ecosystem scientists.

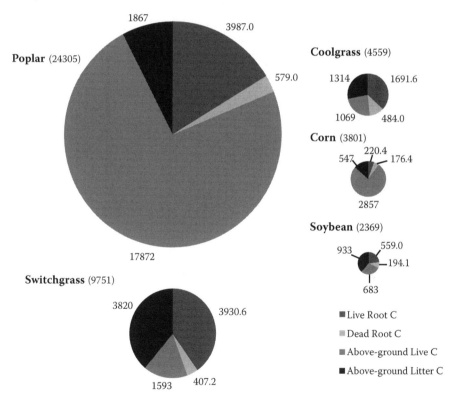

FIGURE 2.2 Carbon partitioning for five species averaged over monthly samples for above-ground components and May–November for below-ground components in a riparian area and adjacent field in central Iowa. Size of each chart indicates relative quantity of carbon measured for each crop. Numbers (kilograms per hectare) represent carbon mass for each component and crop total (in parentheses). (*Source:* Adapted from Tufekcioglu et al., 2003. *Agrofor. Syst.* 57: 187–198.)

COVER CROPS

Cover crops are used for several reasons, primarily for protecting soil by decreasing erosion (Langdale et al., 1991), fixing nitrogen (Blevins et al., 1990), capturing leachable nitrogen from the profile (Singer et al., 2005), and capturing carbon for soil quality maintenance (Ding et al., 2006). Cover crops have been studied for decades (Odland and Knoblauch, 1938; Ebelhar et al., 1986; Singer et al., 2005), but attempts to integrate them broadly into row crop systems, especially in northern regions, have bred limited success (Singer et al., 2007). However, that does not negate their potential positive impact on soil quality if used successfully—particularly their potential positive soil quality impact if integrated with biomass removal schemes. Multiple examples of successful use exist and substantial benefits have been documented under a variety of situations and conditions (Sustainable Agriculture Network, 2007).

Cover crops must not compete intensively with a primary crop, but ideally they will offer living surface cover after the primary crop is harvested and before the next-year crop is established (Sustainable Agriculture Network, 2007). Additionally, if a suitable legume can be used, fixed nitrogen can replace a portion of what would normally be purchased commercially (Blevins et al., 1990). However, a major challenge in successful cover crop use involves establishment of winter covers following primary crop harvest because many primary crops, especially corn, are harvested after a full growing season plus grain dry-down time.

Many farmers delay corn harvest until well after physiologic maturity to facilitate lower grain moisture content, thus reducing grain drying cost. These harvest delays capture valuable solar and thermal resources for cover crop growth. Risk of unfavorable environmental conditions for germination and early growth and risk of spring conditions unfavorable for cover crop management are significant barriers to increased cover crop adoption. Regions such as the southern United States that have longer growing seasons seem to have greater potential for cover crop use, particularly for early harvested crops such as soybean or seed corn, and have better chances of cover crop establishment and early growth success (Singer et al., 2005).

Cover crops typically have shallower roots than primary crops; they should not compete significantly for the same water and nutrient resources required by the primary crop. Cover and primary crops should grow actively at different times of the year. Cover crops contribute root biomass to the soil shallow layers and can contribute substantial amounts of above-ground biomass to the soil surface layer when its growth is terminated. The surface and shallow soil layers that are most deprived of organic materials under biomass harvest schemes are most favorably affected by cover crops. The opportunity to enhance soil ecosystem services through improvement of the surface layer via cover crops seems substantial in the presence of primary crop residue removal.

SYNOPSIS

The cellulosic bioenergy industry may take alternative paths, and each path will affect soil quality differently. One potential pathway involves a business-as-usual scenario that uses the current basic crop production management practices with increased crop residue removal for conversion to liquid fuels. Of the potential pathways, this seems the least desirable relative to soil quality.

Biomass harvest may present opportunities to alter the business-as-usual practices and reduce soil quality impacts resulting from biomass removal. Perlack et al. (2005) assume no-till will be practiced on all biomass harvested fields. This would require considerable changes in many parts of the Corn Belt, for example, where most corn land is tilled. In reality, removal of above-ground corn residue would minimize selected no-till challenges, potentially leading to increased no-till adoption. No-till reduces SOM oxidation compared to oxidation resulting from tillage, potentially offsetting at least partially the impact of biomass removal on SOM loss and soil quality degradation. We do not have a clear understanding of crop residue impacts on surface layer soil quality, i.e., how much residue is required to maintain or improve soil quality of this layer. This is critical because the surface layer regulates flow of water and air to and from the soil profile and plays a critical role in soil warming and cooling.

In concept, cover crops seem excellent matches for row crops undergoing residue removal. Because most SOM originates from root-derived materials, shallow actively growing cover crop root systems may offset SOM losses associated with biomass removal, although this will depend greatly on cover crop growth characteristics (Fronning et al., 2008). Additionally, soil erosion would be minimized by above-ground cover, nitrogen conservation could be significant, and soil quality impacts of fall tillage (if used) would be minimized. The research community's challenge is in developing a row crop–cover crop system that would appeal to large-scale producers. Since corn has been identified as the crop species contributing the most biomass for liquid fuel feedstock (Perlack et al., 2005), development of an acceptable corn–cover cropping system seems paramount.

Soil application of biochar, a thermochemical biomass conversion product, may change soil quality and ecological functions. This carbon-rich material exhibits high specific surface area, high CEC, and low particle density; it offers an effective means of carbon sequestration that can be quantified for carbon trading markets. Its impact on soil quality is under study, and early results indicate it may affect parameters such as water holding capacity, CEC, pH, bulk density, nitrous oxide emissions, and retention of fugitive chemicals. Much work is required to determine if, when, and where biochar applications will positively impact soil quality.

Biomass feedstock conversion platform will influence which cropping pathway we use. The first pathway—focusing on removal of existing crop residues for feedstock, the business-as-usual scenario—will be fostered by a platform that requires a singular or narrow range of species for conversion. Should a conversion platform accept multiple species concurrently, the management scenario could be modified. Perennials or perennial mixes could be converted to cellulosic liquid fuel products with crop biomass from production fields, causing a market signal for increased diversity on farmed lands. Dedicated perennials make the most economic sense on lands offering marginal economic returns for traditional row crops, for example, highly eroded areas, highly erodible lands, areas in fields receiving concentrated water runoff flow with heavy rainfall, and riparian areas subjected to periodic flooding.

Perennials, and especially perennial grasses, seem to maintain or enhance soil quality even with above-ground biomass removal (Liebig, et al., 2008). Enhanced soil quality, especially on degraded land areas, may result from bioenergy activities if the flexible biofuel feedstock conversion platform emerges as the dominant conversion method. However, if we rely on a business-as-usual soil management approach to biofuels, soil quality will very likely undergo additional degradation.

REFERENCES

Al-Durrah, M. and J.M. Bradford. 1981. New methods for studying soil detachment due to waterdrop impact. *Soil Sci. Soc. Am. J.* 45: 949–953.

Allmaras, R.R., D.R. Linden, and C.E. Clapp. 2004. Corn-residue transformations into root and soil carbon as related to nitrogen, tillage and stover management. *Soil Sci. Soc. Am. J.* 68: 1366–1375.

Anderson, S.H., C.J. Gantzer, and J.R. Brown. 1990. Soil physical properties after 100 years of continuous cultivation. *J. Soil Water Conserv.* 45: 117–121.

Baker, J.M. et al. 2006. Tillage and soil carbon sequestration: What do we really know? *Agric. Ecosyst. Env.* 118: 1–5.

Balesdent, J. and M. Balabane. 1992. Maize root-derived soil organic carbon estimated by natural ^{13}C abundance. *Soil Biol. Biochem.* 24: 97–101.

Balesdent, J. and M. Balabane. 1996. Major contribution of roots to soil carbon storage inferred from maize cultivated soils. *Soil Biol. Biochem.* 28: 1261–1263.

Barnhart, S. L., W. D. Schrader, and J. R. Webb. 1978. Comparison of soil properties under continuous corn grain and silage cropping systems. *Agron. J.* 70: 835–837.

Barthès, B. and E. Roose. 2002. Aggregate stability as an indicator of soil susceptibility to runoff and erosion: Validation at several levels. *Catena* 47: 133–149.

Bilbro, J. D. and D.W. Fryrear. 1994. Wind erosion losses as related to plant silhouette and soil cover. *Agron J.* 86: 550–553.

Bissonnias, Y. 1996. Aggregate stability and assessment of soil crustability and erodibility I. Theory and methodology. *Eur. J. Soil Sci.* 4: 435–437.

Blevins, R.L., J.H. Herbek and W.W. Frye. 1990. Legume cover crops as a nitrogen source for no-till corn and grain sorghum. *Agron. J.* 82: 769–772.

Bransby, D.I., McLaughlin, S.B., and Parrish, D.J. 1998. Soil carbon changes and nutrient cycling associated with switchgrass. *Biom. Bioenergy* 14: 379–384.

Bresler, E. and W.D. Kemper. 1970. Soil water evaporation as affected by wetting methods and crust formation. *Soil Sci. Soc. Am. Proc.* 34: 3–8.

Burwell, R.E. and W.E. Larson. 1969. Infiltration as influenced by tillage-induced random roughness and pore space. *Soil Sci. Soc. Amer. Proc.* 33: 449–452.

Cheshire, M.V., G.P. Sparking, and C.M. Mundie. 1983. Effect of periodate treatment of soil on carbohydrate constituents and soil aggregation. *J. Soil Sci.* 34: 105–112.

Council for Agricultural Science and Technology. 2004. Climate change and greenhouse gas mitigation: Challenges and opportunities for agriculture. Task Force Report 141. Ames, IA.

Crovetto Lamarca, C. 1996. Stubble over the soil: Vital role of plant residue in soil management to improve soil quality. American Society for Agronomy, Madison, WI.

Cruse, R.M. and W.E. Larson. 1977. Effect of soil shear strength on soil detachment due to raindrop impact. *Soil Sci. Soc. Am. J.* 41: 777–781.

Cruse, R.M., R. Mier, and C.W. Mize. 2001. Surface residue effects on erosion of thawing soil. *Soil Sci. Soc. Am. J.* 65: 178–184.

Ding, G. et al. 2006. Effect of cover crop management on soil organic matter. *Geoderma* 130: 229–239.

Doran, J.W. et al. 1994. *Defining Soil Quality for a Sustainable Environment.* Special Publication 35, Soil Science Society of America, Madison, WI.

Doran, J.W. and A.J. Jones. 1996. *Methods for Assessing Soil Quality.* Special Publication 49, Soil Science Society of America, Madison, WI.

Duffy, M. and D. Smith. 2008. Estimated cost of crop production in Iowa, 2008. Iowa State University Extension. FM-1712.

Ebelhar, S.A., W.W. Frye, and R.L. Blevins. 1984. Nitrogen from legume cover crops for no-tillage. *Agron. J.* 76: 51–55.

Edwards, W.M., M.J. Shipitalo, and L. B. Owens. 1993. Gas, water and solute transport in soils containing macropores: A review of methodology. *Geoderma* 57: 31–49.

Ellert, B.H. and H.H. Janzen. 1999. Short-term influence of tillage on CO_2 fluxes from a semi-arid soil on the Canadian prairies. *Soil Tillage Res.* 50: 21–32.

Farrell, A.E. et al. 2006. Ethanol can contribute to energy and environmental goals. *Science* 311: 506–508.

Fixen, P.E. 2007. Potential biofuels: Influence on nutrient use and removal in the U.S. *IPNI Better Crops* 91: 12–14.

Francis, P.B. and R.M. Cruse. 1983. Soil water matrix: Potential effects on aggregate stability. *Soil Sci. Soc. Am. J.* 47: 478–481.

Franzluebbers, A.J. 2002. Water infiltration and soil structure related to organic matter and its stratification with depth. *Soil Tillage Res.* 66: 197–205.

Fronning, B.E., K.D. Thelen, and D.H. Min. 2008. Use of manure, compost, and cover crops to supplant crop residue carbon in corn stover removed cropping systems. *Agron. J.* 100: 1703–1710.

Gardner, H.R. and W.R. Gardner. 1969. Relation of water application to evaporation and storage of soil water. *Soil Sci. Soc. Am. Proc.* 33: 192–196.

Goldberg, E.D. 1985. *Black Carbon in the Environment: Properties and Distribution.* John Wiley & Sons, New York.

Green, C.J., A.M. Blackmer, and R. Horton. 1995. Nitrogen effects on conservation of carbon during corn residue decomposition in soil. *Soil Sci. Soc. Am. J.* 59: 453–459.

Hartmann, R., H. Verplancke, and M. De Boodt. 1981. Influence of soil surface structure on infiltration and subsequent evaporation under simulated laboratory conditions. *Soil Tillage Res.* 1: 351–359.

Holmes, J.W., E.L. Graecon, and G.C. Gurr. 1960. The evaporation of water from bare soils with different tilths. *Trans. 7th Int. Soil Sci. Conf.* 1: 188–194.

Hooker, B.A. et al. 2005. Long-term effects of tillage and cornstalk return on soil carbon dynamics. *Soil Sci. Soc. Am. Proc.* 69: 188–196.

Hudson, B.D. 1994. Soil organic matter and available water capacity. *J. Soil Water Conserv.* 49: 189–194.

Idso, S.B. et al. 1974. The three stages of drying of a field soil. *Soil Sci. Soc. Am. Proc.* 38: 831–837.

Johnson J.M.F. et al. 2006a. A matter of balance: Conservation and renewable energy. *J. Soil Water Conserv.* 61: 120A–125A.

Johnson, J.M.F., R.R. Allmaras, and D.C. Reicosky. 2006b. Estimating source carbon from crop residues, roots and rhizodeposits using the National Grain-Yield Database. *Agron. J.* 98: 622–636.

Jones, O.R., V.L. Hauser, and T.W. Popham. 1994. No-tillage effects on infiltration, runoff, and water conservation on dryland. *Trans. ASAE* 37: 473–479.

Kanchikerimath, M. and D. Singh. 2001. Soil organic matter and biological properties after 26 years of maize–wheat–cowpea cropping as affected by manure and fertilization in a cambisol in a semiarid region of India. *Agric. Ecosystems Environ.* 86: 155–162.

Karlen, D.L. et al. 1994. Corn residue effects on soil quality following 10 years of no-till corn. *Soil Tillage Res.* 31: 149–167.

Kaspar, T.C., D.C. Erbach, and R.M. Cruse. 1990. Corn response to seed-row residue removal. *Soil Sci. Soc. Am. J.* 54: 1112–1117.

Kemper, W.D. and D.E. Miller. 1974. Management of crusting soils: Some practical possibilities. In Carry, J.E. and Evans, D.D., Eds., *Soil Crusts Technical Bulletin 214.* Arizona Agriculture Experimental Station, Tucson.

Khan, S.A. et al. 2007. The myth of nitrogen fertilization for soil carbon sequestration. *J. Environ. Qual.* 36: 1821–1832.

Laflen, J.M. and T.S. Colvin. 1981. Effect of crop residue on soil loss from continuous row cropping. *Trans. ASAE* 24: 605–609.

Laird, D.A. 2008. The charcoal vision: A win–win–win scenario for simultaneously producing bioenergy, permanently sequestering carbon, while improving soil and water quality. *Agron. J.* 100: 178–181.

Lal, R. 2000. A modest proposal for the year 2001: We can control greenhouse gases in the world...with proper soil management. *J. Soil Water Conserv.* 55: 429–433.

Lal, R. 2001. World cropland soils as a source or sink for atmospheric carbon. *Adv. Agron.* 71: 145–191.

Lal, R. 2002. Soil carbon dynamics in cropland and range land. *Env. Pollut.* 116: 353–362.

Lal, R. 2004. Soil carbon sequestration impacts on global climate change and food security. *Science* 304: 1623–1627.

Lal, R. 2005. World crop residues production and implications of its use as a biofuel. *Environ. Int.* 31: 575–584.

Lal, R. et al. 1998. *Potential of U.S. Cropland for Carbon Sequestration and Greenhouse Effect Mitigation.* Ann Arbor Press, Chelsea, MI.

Lampurlanés, J. and C. Cantero-Martínez. 2003. Soil bulk density and penetration resistance under different tillage and crop management systems. *Agron J.* 95: 526–536.

Langdale, C.G. et al. 1991. Cover crop effects on soil erosion by wind and water. In Hargrove, W.L., Ed., *Cover Crops for Clean Water.* Soil and Water Conservation Society, Ankeny, IA, pp. 15–23.

Larney, F.J. et al. 1997. Changes in total, mineralizable and light fraction soil organic matter with cropping and tillage intensities in semiarid southern Alberta, Canada. *Soil Till. Res.* 42: 229–240.

Larson, W.E., R.F. Holt, and C.W. Carlson. 1978. Residues for soil conservation. In Oschwald, W.R., Ed., *Crop Residue Management Systems.* Special Publication 31. American Society of Agronomy, Madison, WI, pp. 1–15.

Lehmann, J. et al. 2003. Nutrient availability and leaching in an archaeological Anthrosol and a Ferrasol of the Central Amazon. *Plant Soil* 249: 343–357.

Lehmann, J. et al. 2003. Soil fertility and production potential. In Lehmann, J. et al., Eds., *Dark Earths: Origin, Properties, Management.* Kluwer, Amsterdam, pp. 105–124.

Lemon, E.R. 1956. The potentialities for decreasing soil moisture evaporation loss. *Soil Sci. Soc. Am. Proc.* 20: 120–125.

Lemus, R. and R. Lal. 2005. Bioenergy crops and carbon sequestration. *Crit. Rev. Plant Sci.* 24: 1–21.

Liebig, M.A. et al. 2008. Soil carbon storage by switchgrass grown for bioenergy. *Bioenergy Res.* 1: 215–222.

Mannering, J.V. and L.D. Meyer. 1963. The effects of various rates of surface mulch on infiltration and erosion. *Soil Sci. Soc. Am. J.* 27: 84–86.

Mann, L., V. Tolbert, and J. Cushman. 2002. Potential environmental effects of corn (*Zea mays* L.) stover removal with emphasis on soil organic matter and erosion. *Agric. Ecosystem Environ.* 89: 149–166.

Mikha, M.M. and C.W. Rice. 2004. Tillage and manure effects on soil and aggregate-associated carbon and nitrogen. *Soil Sci. Soc. Am. J.* 68: 809–816.

Moebius-Clune, B.N. et al. 2008. Long-term effects of harvesting maize stover and tillage on soil quality. *Soil Sci. Soc. Am. J.* 72: 960–969.

Odland, T.E. and H.C. Knoblauch. 1938. The value of cover crops in continuous corn culture. *J. Am. Soc. Agron.* 30: 22–29.

Olness, A. and D. Archer. 2005. Effect of organic carbon on available water in soil. *Soil Sci.* 170: 90–101.

Perlack, R.D. et al. 2005 Biomass as Feedstock for a Bioenergy and Bioproducts Industry: The Technical Feasibility of a Billion-Ton Annual Supply. U.S. Department of Energy and U.S. Department of Agriculture. http://www.eere.energy.gov/biomass/pdfs/final_billionton_vision_report2.pdf (verified March 1, 2007).

Reicosky, D.C. 1997. Tillage-induced CO_2 emission from soil. *Nutr. Cycling Agroecosyst.* 49: 273–285.

Reicosky, D.C. and D.W. Archer. 2007. Moldboard plow tillage depth and short-term carbon dioxide release. *Soil Tillage Res.* 94: 109–121.

Reicosky, D.C. and M.J. Lindstrom. 1993. Fall tillage method: Effect on short-term carbon dioxide flux from soil. *Agron. J.* 85: 1237–1243.

Reicosky, D.C. et al. 2002. Continuous corn with moldboard tillage: Residue and fertility effects on soil carbon. *J. Soil Water Conserv.* 57: 277–284.

Reicosky, D.C. and A.R. Wilts. 2004. Crop-residue management. In Hillel, D., Ed., *Encyclopedia of Soils in the Environment*, Vol. 1. Elsevier, Oxford, pp. 334–338.

Robert, M. 1996. Aluminum toxicity a major stress for microbes in the environment. In Huang, P.M. et al., Eds., *Environmental Impacts, Vol. 2: Soil Component Interactions*. CRC Press, Boca Raton, FL, pp. 227–242.

Robert, M. 2001. Carbon Sequestration in Soils: Proposals for Land Management, Report 96. United Nations Food and Agricultural Organization, Rome.

Robinson, C.A., Cruse, R.M., and Ghaffarzadeh, M. 1996. Cropping systems and nitrogen effects on mollisol organic carbon. *Soil Sci. Soc. Am. J.* 60: 264–269.

Russell, A.E. et al. 2005. Impact of nitrogen fertilizer and cropping system on carbon sequestration in Midwestern mollisols. *Soil Sci. Soc. Am. J.* 69: 413–422.

Schmidt, M.W.I. and A.G. Noack. 2000. Black carbon in soils and sediments: Analysis, distribution, implications, and current challenges. *Global Biogeochem. Cycles*, 14: 777–793.

Singer, J., T. Kaspar, and P. Pederson. 2005. Small grain cover crops for corn and soybean. Iowa State University Extension Program.

Singer, J.W., S.M. Nusser, and C.J. Alf. 2007. Are cover crops being used in U.S. Corn Belt? *J. Soil Water Conserv.* 62: 353–358.

Six, J. et al. 2000. Soil structure and organic matter I: Distribution of aggregate-size classes and aggregate-associated carbon. *Soil Sci. Soc. of Am. J.* 64: 681–689.

Soane, B.D. 1990. The role of organic matter in soil compactibility: A review of some practical aspects. *Soil Tillage Res.* 16: 179–201.

Sojka, R.E. and D.R. Upchurch. 1999. Reservations regarding the soil quality concept. *Soil Sci. Soc. Am. J.* 63: 1039–1054.

Sommerfeldt, T.E., C. Chang, and T. Entz. 1988. Long-term annual manure applications increase soil organic matter and nitrogen, and decrease carbon to nitrogen ratio. *Soil Sci. Soc. Am. J.* 52: 1668–1672.

Stockfisch, N., T. Forstreuter, and W. Ehlers. 1999. Ploughing effects on soil organic matter after 20 years of conservation tillage in Lower Saxony, Germany. *Soil Tillage Res.* 52: 91–101.

Sustainable Agriculture Network. 2007. *Managing Cover Crops Profitably*, 3rd ed. Handbook Series, Book 9.

Tebrugge, F. and R.A. Guring. 1999. Reducing tillage intensity: Review of results from a long-term study in Germany. *Soil Tillage Res.* 53: 15–28.

Tisdall, J.M. and J.M. Oades. 1982. Organic matter and water-stable aggregates in soils. *Eur. J. Soil Sci.* 33: 141–163.

Tufekcioglu, A. et al. 2003. Biomass, carbon and nitrogen dynamics of multi-species riparian buffers within an agricultural watershed in Iowa, USA. *Agrofor. Syst.* 57: 187–198.

Unger, P.W. 1994. *Managing Agricultural Residues*, CRC Press, Boca Raton, FL.

Uri, N.D. 1999. Conservation tillage in U.S. agriculture. In *Environmental, Economic, and Policy Issues.* Haworth Press, Binghamton, NY, p. 130.

Vitosh, M.L., R.E. Lucas, and G.H. Silva. 1997. Long-term effects of fertilizer and manure on corn yield, soil carbon and other soil chemical properties in Michigan. In Paul, E.A. et al., Eds., *Soil Organic Matter in Temperate Agroecosystems: Long-term Experiments in North America.* CRC Press, Boca Raton, FL, pp. 129–139.

Wardle, D.A., M.C. Nilsson, and O. Zackrisson. 2008. Fire-derived charcoal causes loss of forest humus. *Science* 320: 629.

Wiesenthal, T. et al. 2006. How much bioenergy can Europe produce without harming the environment? European Environmental Agency Report 7/2006. Luxembourg.

Wilhelm, W.W. et al. 2007. Corn stover to sustain soil organic carbon further constrains biomass supply. *Agron. J.* 99: 1665–1667.

Wilhelm, W.W. et al. 2004. Crop and soil productivity response to corn residue removal: A literature review. *Agron. J.* 96: 1–17.

Williams, B.B. et al. 1966. Techniques for the determination of the stability of soil aggregates. *Soil Sci.* 101: 157–163.

Willis, W.O. and J.J. Bond. 1971. Soil water evaporation: Reduction by simulated tillage. *Soil Sci. Soc. Am. Proc.* 35: 526529.

Zan, C.S. et al. 2001. Carbon sequestration in perennial bioenergy, annual corn and uncultivated systems in southern Quebec. *Agric., Ecosyst. Environ.* 86: 135–144.

3 Ecological Consequences of Biofuels

W.E.H. Blum, M.H. Gerzabek, K. Hackländer,
R. Horn, F. Reimoser, W. Winiwarter,
S. Zechmeister-Boltenstern, and F. Zehetner

CONTENTS

INTRODUCTION AND BACKGROUND

Soils provide important goods and services to humankind and the environment (Blum, 2005); both are increasingly under threat due to global development (Crutzen, 2002). For the first time in recent history, world food production proved insufficient for sustaining human societies at affordable prices in 2007 and 2008. One reason is the decrease of soil surfaces for food and fiber production arising from soil losses by sealing and further forms of degradation because only 12% of the global land surface is suitable for food and fiber production. Of the remainder, 24% is for grazing, 31% is forest, and 33% is unsuitable for any use (Blum and Eswaran, 2004). This means that 12% of the land surface and about 25% of the world population produce all worldwide traded foodstuffs, predominantly in the northern hemisphere. What are the future perspectives?

As our ancestors started to settle on fertile soils, urbanization occurred worldwide on the most fertile land and continues today. A rough estimate for Europe reveals that 9 to 10 km² of fertile soil daily are used for the construction of traffic ways, settlements, industrial premises, and other urban purposes. A list of areas of land degradation by country based on 23-year global inventory, modeling, and mapping studies (Bai et al., 2008) indicates that about one-fourth of the world soils are severely degraded. A regional breakdown of total human appropriation of net primary production including wood biomass shows 50% appropriation in some parts of the world—for example, 63% in Southern Asia, 52% in Eastern and Southeastern Europe, and 40% and 35%, respectively, in Western Europe and Eastern Asia (Haberl et al., 2007). Losses of productivity compared to potential vegetation globally indicate that current vegetation fixes approximately 10% less organic carbon compared to potential vegetation (Haberl et al., 2007). In 2000, 16% of the global terrestrial net primary production was already used by humans: 12% serving as food, 58% as feed for livestock, 20% as raw material, and 10% as fuel wood (Krausmann et al., 2008). This intensive land use constitutes one of the important global pressures on biodiversity and natural resources.

The world population continues to increase by about 85 million annually, and within the next decade about 1 billion people will move from rural to urban environments, thus destroying nutritional self-sufficiency and depending on local, regional, and world food markets. Moreover, the demand for animal protein is exponentially increasing, and this creates demands for enormous amounts of basic staple foods. These are clear indicators that we are reaching threshold conditions in biomass production for worldwide supplies of food and feed.

Lal (2006) estimated that average grain yields must be improved from 2.64 Mg ha^{-1} (in 2000) to approximately 4.3 Mg ha^{-1} by 2050 to cover the needs of a world population of 10 billion people. How can this goal be reached in view of a completely new production line that increasingly uses agricultural land formerly dedicated to food production for production of biofuels to replace fossil fuels? In addition to the issue of competition of biofuel and food production for the same agricultural land, we must also ask what the ecological effects of biofuel production may be in the medium to long run.

The European Environment Agency (EEA) estimates a fourfold increase of the bioenergy potential of the European Union (EU) between 2006 and 2030, based mainly on agricultural production (2006). The benefits of biofuel plants for CO_2 and the use of biomass to generate electricity are under debate (Powlson et al., 2005). However, the expected magnitude of land use changes due to biofuel cropping is tremendous, and the impact of biofuel cropping on ecosystems and the environment will be diverse.

Perennial bioenergy cropping schemes such as the switchgrass (*Panicum virgatum L.*) system introduced in the United States and the hybrid poplar may be beneficial because they reduce soil erosion, enhance soil carbon sequestration, improve wildlife habitats, and reduce greenhouse gas emissions (Sanderson et al., 2006). Annual systems, however, may exert additional pressures on the environment. Biofuels of the first generation (e.g., bioethanol from cereal grains or biodiesel from oil seeds) require intensive inputs such as N fertilizers, and may become large monocultures posing additional problems from heavy machinery and subsequent soil compaction. Soil biodiversity and plant and wildlife diversity will also be subject to influence. Biofuel production based on the cellulose-to-liquid system to generate the so-called second generation biofuels may lead to further soil organic matter decline if aboveground plant residues are removed completely. However, based on our current state of knowledge of the impacts of biofuel production on agricultural land and the increasing pace of biofuel production worldwide, few measured data are available to allow sound judgments about the ecological impacts of biofuel production on soil. Nevertheless, this chapter will highlight some of the possible impacts with a special focus on ecological soil functions and biodiversity.

BIOFUEL CROPPING AND SOIL ORGANIC MATTER

INTRODUCTION

Soil organic matter (SOM) management is one important option to mitigate adverse impacts on soil resources and sustain soil functions. The level and quality of SOM in soils is mainly influenced by levels of SOM input (Gerzabek et al., 1997), tillage management (Alvarez, 2005), possible topsoil losses and mineralization induced by erosion (Polyakov and Lal, 2008), soil moisture and temperature effects on microbial activity, and the types and amounts of stabilizing inorganic constituents such as Fe oxides and clay minerals that provide reactive surfaces for interactions with humified material (Kaiser et al., 2007). Whether SOM levels will decrease in response to climate change due to enhanced mineralization or increase because of enhanced net primary production and plant residue input is an ongoing debate. Clearly, SOM is sensitive to temperature changes and the effects of climatic regimes on SOM decomposition are well-documented (Couteaux et al., 2001; Emmett et al., 2004). However, many other environmental constraints such as substrate availability and stability and soil moisture must be considered in calculating future scenarios of SOM levels and dynamics (Davidson and Janssens, 2006).

Comparing the impacts of various land use and management options, we observe different responses of soil organic carbon (SOC) to practical measures (Gerzabek, 2007). The most important factor governing SOC levels at a site is agricultural or silvicultural land use. In Austria, median SOC stocks (0 to 50 cm depth) in extensively used grasslands are twice as high as those in crop lands (Gerzabek et al., 2005). Fertilizer regime is another important factor. In a long-term experiment in Sweden, peat treatment produced SOC content more than three times higher than black fallow after about 40 years. Likewise, animal manure treatment produced a 1.5 times higher SOC content in the 0 to 20 cm layer compared to N fertilizer mineral treatment (Kirchmann et al., 2004). Comparatively little variation in SOC stocks resulted from tillage management. In a long-term experiment in Austria, black fallow without tillage showed a 1.11 times higher SOC content than cropped and tilled plots after 38 years (Antil et al., 2005).

SOM is one of the most important properties for preserving the integrity of ecological soil functions such as the buffer and filter activities, habitat maintenance, and production (Arshad and Martin, 2002; Carter, 2002). Therefore, we have chosen SOM as an indicator of possible effects of biofuel cropping on soil functions. In the following sections, we assess the impacts of biofuel cropping on SOM dynamics by simulation modeling based on three long-term experiments in the United Kingdom, Sweden, and Austria. The specific research issue was the long-term effect of residue removal on SOM levels under different soil conditions and climatic regimes.

SIMULATION MODELING

The three long-term experiments took place in Rothamsted, U.K., Ultuna, Sweden, and Fuchsenbigl, Austria, in 1888, 1956, and 1967, respectively, and involve Luvisols (Alfisols; Rothamsted), Cambisols (Inceptisols; Ultuna), and Chernozems (Mollisols; Fuchsenbigl). Figure 3.1 shows average monthly mean temperature and precipitation at the three experimental sites. Detailed descriptions of the field experiments can be found in Jenkinson et al., 1992; Gerzabek et al., 1997; and Oberländer and Roth, 1974, respectively.

We used the experimental data to calibrate the carbon decomposition rate constants of the Rothamsted Carbon Model (version RothC-26.3) for the three experimental sites. The model allows calculation of the effects of organic matter management on the development of SOC in non-waterlogged top soils over intervals ranging from a few years to a few centuries. In the model, SOC is divided into four active compartments and a small portion of *inert organic matter* (IOM). The four active compartments are *decomposable plant material* (DPM), *resistant plant material* (RPM), *microbial biomass* (BIO), and *humified organic matter* (HUM). Each compartment decomposes by first-order kinetics at its own characteristic rate. The IOM compartment is considered resistant to decomposition. RothC-26.3 is based on a monthly time step calculation (Jenkinson and Coleman, 1994).

The model requires certain input data: (1) climate data (average monthly mean temperatures, precipitation, and evapotranspiration); (2) land management data (monthly input of plant residues, farmyard manure, and soil cover data); (3) SOC and clay content; and (4) sampling depth and partitioning of SOC among the different carbon pools.

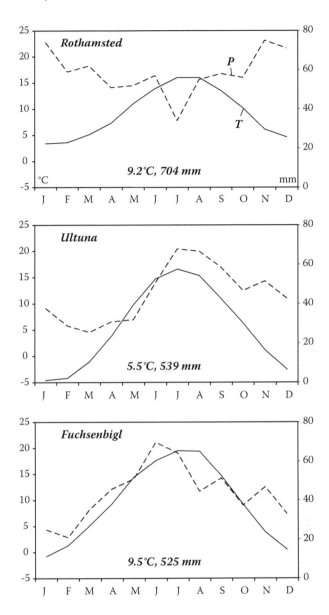

FIGURE 3.1 Climatic parameters at the three experimental sites.

The annual input of organic C to the soil is difficult to measure because it derives from non-harvested plant parts like roots and stubbles, litter fall, root exudates, etc. (Jenkinson et al., 1992). To calculate plant inputs to the soil: (1) the ratio of DPM to RPM was set to 1.44 as suggested by Jenkinson et al. (1992); (2) the partitioning of crop biomass into roots-plus-chaff, straw, and grain compartments was estimated according to Aufhammer (1998; grain maize, winter wheat, spring barley) and Malhi et al. (2006; oilseed rape); (3) approximately 44% of the total plant residues entering the soil were considered to be organic carbon (McKendry, 2002); (4) half of the annual plant input to the soil was attributed to the harvesting month, and the other half was distributed within the vegetation period starting in April, as suggested by Jenkinson et al. (1992); and (5) the IOM pool was calculated as IOM = 0.049 × $SOC^{1.139}$ (Fallon et al., 1998).

We used the calibrated RothC-26.3 model to assess the long-term impact of straw removal from biofuel cropping on SOC stocks at the three experimental sites. In agronomic practice, biofuel crops are generally included in crop rotations. However, for the sake of better comparison of biofuel crops, we simulated continuous cropping of grain maize, winter wheat, spring barley, and oilseed rape, respectively. Climate input data were based on past records and plant residue input was estimated from average crop yields of the three experimental regions.

IMPACTS OF BIOFUEL CROPPING ON SOM LEVELS IN DIFFERENT CLIMATIC REGIONS

The modeling outputs are presented in Figure 3.2 (grain maize) and Table 3.1 (winter wheat, spring barley, and oilseed rape). Complete removal of the maize stalks and cobs (maize stover) resulted in substantial SOC declines at the Rothamsted and Ultuna sites, but had little impact at Fuchsenbigl (Figure 3.2). In the Chernozems at Fuchsenbigl, the HUM pool decomposed at a relatively slow rate (0.005 yr^{-1}), probably caused by the formation of stable organo-mineral complexes in this very biologically active soil. Organo-mineral complexes are more resistant to microbial attack than uncomplexed organic materials (Quideau et al., 2000) and so favor the build up of SOM. Our results further demonstrate that if 50% of the maize stover was left in the field, SOC stocks showed increasing trends over time at all three sites (Figure 3.2).

The modeling results for winter wheat and spring barley indicated slight SOC decreases at the Fuchsenbigl site when 100% of the straw was removed and increasing trends when 50% was removed. However, at the Rothamsted and Ultuna sites, 50% straw removal still resulted in declining SOC (Table 3.1). By contrast, oilseed rape increased SOC stocks at all sites due to its high root biomass, even when 100% straw was removed. Our results demonstrate that the removal of crop residues, for example, to produce cellulosic ethanol, may entail a long-term decline of SOM. This result is obviously unsustainable; it compromises soil functional integrity and may trigger adverse effects such as decreased fertility and increased runoff and erosion. However, these impacts strongly depend on crop type, soil properties, and climatic conditions.

POSSIBLE LONG-TERM EFFECTS OF CONTINUOUS BIOMASS PRODUCTION AND HARVESTING ON SOIL PROPERTIES

INTRODUCTION

Stress applied to soil surfaces during wheeling or harvesting is transmitted three dimensionally and alters pore size distribution and pore functioning, including physico-chemical and biological properties, as soon as the internal soil strength is exceeded. Processes of soil deformation, including compaction and shear effects, are

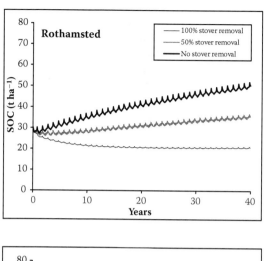

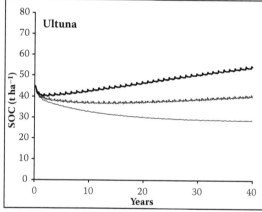

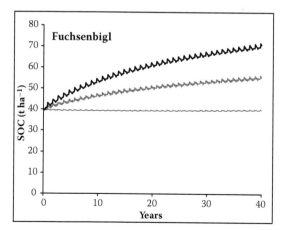

FIGURE 3.2 Modeling of soil organic carbon stocks (0 to 20 cm depth) at the three experimental sites with maize and different levels of stover removal.

TABLE 3.1

Modeling of Soil Organic Carbon Stocks (0 to 20 cm depth) at Different Levels of Straw Removal at Experimental Sites

SOC stocks in 0 to 20 cm depth (t ha⁻¹)

	Start Value	Winter Wheat			Spring Barley			Oilseed Rape		
		End Value after 40 Years			End Value after 40 Years			End Value after 40 Years		
		100% Straw Removal	50% Straw Removal	No Straw Removal	100% Straw Removal	50% Straw Removal	No Straw Removal	100% Straw Removal	50% Straw Removal	No Straw Removal
Rothamsted	30.4	19.1	24.8	30.3	17.5	21.8	25.8	34.4	40.9	47.3
Ultuna	45.0	27.9	38.5	49.1	38.3	43.0	47.6	47.3	55.5	63.7
Fuchsenbigl	39.8	39.0	44.8	50.7	37.2	41.5	46.1	54.2	60.9	67.6

well-understood and documented, especially for arable and forest soils and to a certain extent for grassland soils (Horn et al., 2000; Pagliai and Jones, 2002; Horn et al., 2006; Wiermann, 1998; Fazekas, 2005; Vossbrink and Horn, 2004; Birkas, 2008).

However, until now, the consequences of harvesting with heavy machines under very unfavorable wet soil conditions (maize for bioenergy production, continuous cutting and harvesting of fast growing willows) have been investigated only marginally with respect to physical and mechanical interactions and resulting changes in depth-dependent soil properties under specific climate and site conditions. Therefore, general background information about the effects of stress strain and hydraulic processes will be discussed on the basis of a detailed initial study of willow harvesting with regard to changes in soil properties.

STRESS STRAIN EFFECTS: GENERAL INFORMATION

The internal strength (precompression stress value as reaction force) for each soil horizon defines the maximal applicable external stress (as action force) that can be withstood without changes in soil physical or physicochemical properties. Soils are weaker and more vulnerable to deformation based on how wet they are. Their properties are also altered (Horn, 1988) if:

- Clay and silt contents of soil structure are higher.
- Aggregates are less rigid.
- Monovalent ions at exchange positions dominate.
- Amount of organic carbon is small.
- Proportion of shear on total forces is increased.

Additionally, under such unfavorable conditions the existing soil structure will be more easily disturbed due to kneading and further compressed, leading to reduced soil aeration and higher emissions of CO_2, N_2O, and even CH_4 to the atmosphere (van den Akker et al., 1995).

Stress-induced strain processes such as those due to wheeling are always time- and frequency-dependent and result in an additive height change and induced reduction in pore volume and functioning as soon as the internal soil strength per soil horizon under given conditions is exceeded. These changes are irreversible if no external reloosening forces are applied via plowing.

It is well known within a given contact area increasing stresses are transmitted to deeper depths; this applies also to a given stress in an increasing contact area. This effect underlines the detrimental effects of the increased use of agricultural and forest machinery (threefold to fourfold in the past 40 years). Furthermore, shear induced soil deformation causes an even more pronounced deterioration of existing soil aggregation compared to simple static stress application and leads to a more horizontal rearrangement of soil particles and formation of platy structures (Peth and Horn, 2006).

However, it has also been proven that an existing platy structure in a plow pan of arable soils, which is in equilibrium with external forces, reacts with a more pronounced strain and reduction in pore volume and functioning when exposed to a stress that exceeds all earlier stresses. If the soil strength in the deeper soil horizons is very small, a complete deterioration of the existing structure at a deeper soil depth may occur (Fazekas, 2005). If the water content during deformation is high and the matric potential values are less negative, soils are deformed even more intensively, because stress application and strain together induce a greater water saturation via decreases of pore volume and coarse pore sizes. If even positive matric potential values occur and if the hydraulic conductivity and hydraulic gradient are not sufficient to drain off excess soil water, soils become more sensitive to deformation (Fazekas and Horn, 2005). These processes are more pronounced in soils when internal soil strength is exceeded due to loading (Horn et al., 2002).

Semmel and Horn (1995) analyzed the effects of 50 wheeling events within a day in a haplic luvisol derived from loess at a given wheel load of 4 Mg at field capacity. They proved that based on the changes in stress components ($\sigma_1, \sigma_2 = \sigma_3$), the vertical major stress σ_1 (defined as a soil weakening) increased relatively with wheeling frequency. The actual matric potential values became less negative and even became positive values attributable to a further strength decline and complete deterioration of existing soil structure (Fazekas, 2005).

Zink et al. (2008) proved that a wheel load of 6.5 Mg resulted in an intense increase in precompression stress down to 60 cm after 10 consecutive wheeling events, while the corresponding wheel load of 7.5 Mg increased both depth and magnitude values further in a Luvisol derived from loess at field capacity. Because the soil strength was smaller than the applied stresses, air permeability and saturated hydraulic conductivity decreased up to three orders of magnitude down to depths of 50 to 60 cm due to the 10 wheeling events within a day. Consequently, we must conclude that if applied stresses exceed the internal strength no reloosening occurs, soil physical and physico-chemical properties deteriorate and do not show significant improvement over time (Horn, 2004). Thus, soils reveal a memory effect because they accumulate even short-term wheeling-induced strain effects. These results agree with those of Wiermann, 1998; Wiermann et al., 2000; and Peng and Horn, 2008, even if variations in soil management (conservation versus conventional tillage) resulted in slightly higher soil strength and better stress attenuation in the conservation-tilled Luvisols derived from loess.

However, even if rigid soils exhibit higher strength values than weak ones, they undergo an identical extra soil deformation in the virgin compression load range, leading to a further and mostly irreversible decline of hydraulic or pneumatic function and corresponding physico-chemical processes such as redox potential and microorganism activities and compositions to greater depths. Schäffer et al. (2001) analyzed stress-induced changes in CO_2 emissions under wheel tracks and concluded that directly under the tracks the CO_2 concentration in the air-filled pores may have exceeded 5% down to 30 cm depth; within a distance of 1 dm from the track, CO_2 concentrations were even higher.

Schrader and Bayer (2000) analyzed changes in microbial biomass and activity due to load application and wheeling and concluded that as soon as the internal soil strength was exceeded by the applied soil stress, the compositions and activities of soil fauna and flora were affected and the effects were more pronounced for continuously plowed sites in comparison to conservation-tilled sites.

EFFECTS OF WILLOW TREE HARVESTING ON SOIL STRENGTH AND AIR PERMEABILITY

The consequences of repeated harvesting of willow plantations for bioenergy production over the years and under unfavorable soil conditions on silty eutric fluvisol derived from marine sediments in northern Germany are documented in Figure 3.3. In addition to a decline in total pore volume up to 10%, complete deterioration of the coarse pores and an increase in the finer pores were detected down to 60 cm depth (not shown).

Repeated wheeling with harvesting equipment (total mass of bunker wagon 12 Mg, two axles) increased the precompression stress values of topsoil and subsoil down to 60 cm depth by a factor of two to three compared to the original values. These increased values underline the enormous effect of additive subsoil compaction even after only three years of tree growth. To estimate the long-term productivity of such plantations, we must critically analyze in more detail the induced changes in air permeability as an example for soil productivity and functions that underwent significant declines even after this short period—about half an order of magnitude compared with unstressed field properties (Piepenbrock, 2002). We can expect changes to be even more pronounced if soil preparation and harvesting processes must be carried out under wet soil conditions in late autumn or early winter and irreversible soil deformation processes comparable to those of sugar beet or maize harvesting campaigns in late autumn under less favorable hydraulic soil conditions can be expected.

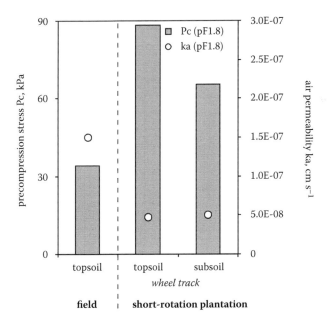

FIGURE 3.3 Effects of willow harvesting on soil strength (= precompression stress) and air permeability in topsoil and subsoil of silty eutric fluvisol derived from marine sediments.

NITROGEN CYCLE AND N$_2$O EMISSIONS

NATURAL CYCLING OF NITROGEN

Nitrogen (N) is an essential element for plant growth. While nitrogen is abundant in the atmosphere (mainly N$_2$ at 78 vol% representing >99% of available nitrogen), it is not accessible to plants. N$_2$ is a very stable molecule that requires considerable energy for conversion into compounds that can be used by plants.

In the preindustrial world, the first mechanism to "fix" atmospheric N$_2$ was lightning. The high temperatures in the plasma and the rapid cooling of gases arising from lightning caused the formation of NO that could be converted to nitric acid and nitrates in the atmosphere. Another and more important process relied on the abilities of certain strains of bacteria, most prominently Rhizobia, to produce enzymes that convert atmospheric N$_2$ into ammonia. The root nodules of leguminous plants provide symbiotic conditions to these bacteria. Legumes thus are able to harvest nitrogen fixed by their symbionts. Galloway et al. (2004) estimated the globally fixed nitrogen from lightning at 5.4 Tg N, while terrestrial biological nitrogen fixation (BNF) contributed 120 Tg N annually in preindustrial times. Another 121 Tg N was assumed to derive from marine BNF.

The poor accessibility of N led plants to develop strategies to make best use of N-containing tissues. Proteins are concentrated in biomass needed for metabolism and growth, and are removed when biomass is used for structural purposes (woody biomass) or before abscission (leaves). For this reason, wood has lower N content than other plant tissues.

In contrast, the cycle of N in soils is "leaky." Soil organic N decomposes to ammonia (mineralization); soil microbes convert ammonia to nitrate (nitrification) or further utilize nitrate for conversion into atmospheric N$_2$ (denitrification). The latter process, while yielding energy to the microbes, causes fixed N to be lost to the atmosphere again. As a side effect of both nitrification and denitrification, N$_2$O, a stable and long-lived trace constituent of atmosphere is released. N$_2$O absorbs infrared radiation and thus acts as a greenhouse gas; its global warming potential has been described as 296 times that of CO$_2$ (Prather et al., 2001).

ANTHROPOGENIC INTERFERENCE

N cycling in the environment has been strongly affected by human behavior. After the importance of N to plant growth was recognized, non-manure applications of N to soils started. Cultivation of legumes for BNF provided important inputs (globally, 40 Tg N per year as of 2005; Galloway et al., 2008). The most striking factor is application of mineral fertilizer (121 Tg N), a result of worldwide availability of industrial ammonia production from the Haber-Bosch process made possible only by cheap fossil energy. An unintentional result of N fixation is combustion of fossil fuels (air pollution of ~25 Tg N annually). This process also returns N to soil, mostly via nitrate deposition.

Ammonia is applied directly to soil as a fertilizer (in solid form as urea or ammonium carbonate; both substances are very volatile and lead to large losses of ammonia to the atmosphere) or oxidized to nitric acid and applied as a nitrate. In both forms, mineral N participates in the natural N cycle of nitrification, denitrification, and plant assimilation. Increasing amounts of fixed N have been applied in excess to soils to the extent that they leak into the environment. An array of environmental problems have resulted from N pollution: soil acidification, eutrophication of surface waters, nitrate pollution of drinking water, release of ammonia to the atmosphere, leading to particle formation (air pollution), atmospheric NO_2 and ozone formation via NO release from soils. Ozone and directly released N_2O are efficient greenhouse gases (GHGs). N is transferred from one environmental pool to the next, creating problems in each pool until it is ultimately released back to the atmosphere as N_2 (mostly as a product of denitrification). As noted above, even the environmentally "benign" process of removing N from the cascade serves merely as a waste of energy because it requires new N to be brought to agricultural land.

Emission of N_2O from soils—only one of many problems associated with N surplus—has been quantified from a multitude of plot scale measurements, mostly from European and North American field sites (Bouwman et al., 2002). Emissions occur episodically and are difficult to measure; their dependence on soil parameters is not fully understood. N_2O formed in the soil is prone to denitrification as it moves toward the surface—a removal process that is very difficult to predict based on soil properties and environmental conditions. IPCC (2006) proposes a release rate based on the extent of N added to soil, suggesting that 1% of each such addition will be converted to N_2O (regardless of origins of N-containing substrates). Questions of limited applicability of this approach are addressed by an uncertainty range that covers an order of magnitude.

Mineral fertilizers obviously are required to provide the world with food. Smil (2001) estimated that as early as the 1990s about 40% of the world population depended on food grown on mineral fertilizers. Independent of any consideration of sustainability, this route of N fixation cannot be reversed without major severe effects on humanity as a whole.

BIOFUELS AND N_2O EMISSIONS

While all agricultural production based on increased N turnaround will inevitably contribute to N_2O emissions, N turnaround related to biofuel production will be critical as long as biofuels are grown with the intent of reducing GHG emissions. N_2O emission is now an element considered in life cycle assessment (LCA), but a clear guideline on the best way to consider this factor is still lacking. In addition to the uncertainty of the IPCC methodology, some of the N_2O release steps cited may not have been fully considered. This is the case for LCAs that fail to cover the indirect emissions specified as such by IPCC (2006).

Crutzen et al. (2008) attempted to estimate emissions as seen from the atmosphere. The pathway and rate of N_2O decay in the stratosphere are fairly well-known. Thus, a top-down assessment will provide more precise results of total N_2O flux into the atmosphere than any approach based on individual release processes. The results of Crutzen et al. (2008) are "not inconsistent" with estimates according to the IPCC approach.

Assuming biofuels are produced in the same manner as conventional crops, Crutzen et al. (2008) determined the need for adding fertilizer N from plant N

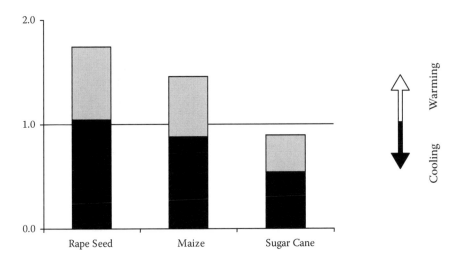

FIGURE 3.4   GHG emission index of biofuels according to Crutzen et al. (2008; default assumptions). Shaded area represents uncertainty range. Biofuels provide advantages in terms of GHG emissions with respect to fossil fuels only for values below unity (cooling).

content, then related the associated N_2O emissions to the fossil fuel replaced and resulting CO_2 emissions. This ratio is larger than unity for the most popular transport biofuels, i.e., emissions of GHGs associated with biofuels are higher than those of fossil fuels (see Figure 3.4).

In a sensitivity analysis, the same authors point to the factors and assumptions that most strongly affected the results. One is nitrogen use efficiency (NUE): the fraction of applied N incorporated in plants. Globally, NUE is only about 40%; the remainder of applied N follows other pathways in the N cascade (emissions as gaseous NH_3, N_2 release after denitrification, leaching of nitrate to groundwater). Field tests on experimental plots showed much higher efficiencies. A realistic target for improvement seems to be a 60% NUE, considering that this factor also covers losses due to crop failure, difficulties in work organization, and inappropriate N application—factors not considered at test plots.

Another critical factor is the extent of replacement of other agricultural production due to use of by-products of biofuel production. Biofuels contain very little N and residues from their production qualify as protein-rich animal fodders. This is the case for production of ethanol (protein in distillation residue) and biodiesel in which press cakes of oilseeds are especially rich in proteins. To the extent these by-products replace other agricultural products and their respective fertilizer use and associated N_2O emissions, the GHG balance of biofuels will improve. Such assumptions drive overall efficiency. We may safely assume that by-products of adequate price and quality will eventually be used, even if they are not yet used in current biofuel production. It is not clear, however, whether optimization of biofuel production will compromise fodder quality, and how the trend of segregation of animal husbandry from land cultivation will affect transport costs.

FUTURE IMPACTS OF SECOND GENERATION BIOFUELS

Future biofuels are expected to be produced from cellulose and lignin; the processes will be able to utilize most of the plants grown for their production (fast-growing woody biomass or grasses like miscanthus). Therefore the area-based yield of second generation biofuels is expected to be considerably higher than that of current production, and will not rely on finding and maintaining adequate use of by-products. Moreover, the target compounds contain little nitrogen, and thus fertilizer rates can be kept low. Forests currently receive no additional fertilization (other than atmospheric deposition), and high yields have been reported for switchgrass without addition of nitrogen (Tilman et al., 2006).

In implementing second generation biofuel production, a few problems still require resolution. They relate to efficient conversion of biomass to biofuels and locating appropriate areas for cultivation since production obviously is not desirable for current agricultural sites best suited for farming that continue to produce food for the growing world population. It remains to be seen whether optimizing area yields would imply, as in classical agriculture, overfertilization to an extent that offsets some of the expected improvements in the GHG balance.

POSSIBLE IMPACTS ON SOIL ORGANISMS

INTRODUCTION

Soil organisms are vital for the functioning of terrestrial ecosystems; they serve as motors of decomposition, nutrient cycling, soil formation, soil structuring, and carbon storage. Some soil microbes can fix nitrogen or remediate pollutants; some play major roles in controlling GHG fluxes. Soil harbors a biological diversity greater than any natural above-ground system and hence acts as a huge gene reserve for many biotechnological applications, particularly for the conversion of biomass to biofuels. A fertile resource for data mining to search for new examples of relevant enzymes has emerged in the form of new metagenomic approaches that eliminate the difficult and time-consuming culture of soil microbial communities. Presently many soil bacteria and fungi are screened and tested for their abilities of cellulytic conversion. Examples include *Fibrobacter succinogenes* (Lynd et al., 2002) and *Cytophaga hutchinsonii*, an aerobic Gram-negative bacterium commonly found in soil that rapidly digests crystalline cellulose (U.S. DOE 2006). These examples illustrate how important it is to protect soils and their rich living communities.

Microbial biomass in the soil and the organic carbon content are closely related. The proportion of microbial biomass carbon has been found to be 1% to 3% of organic carbon (Sparling, 1985). While microbes constitute the basis of the soil–food web, all soil organisms depend strongly on soil carbon storage. When comparing different production systems (crops, grasslands, plantations, and native forests) for biofuels, land use change is a major factor determining impacts on soil in general and soil organisms in particular.

Guo and Gifford (2002) conducted a meta-analysis of the effects of land use change on soil carbon. Their review revealed that soil C stocks tend to decline after land use changes from pasture to plantation (–10%), native forest to plantation (–13%), native forest to crop (–42%), and pasture to crop (–59%). Soil C stocks increase after land use changes from native forest to pasture (+8%), crop to pasture (+19%) crop to plantation (+18%), and crop to secondary forest (+53%). Broadleaf tree plantations placed onto prior native forests and pastures did not affect soil C stocks; pine plantations reduced soil C stocks by 12% to 15%.

Land use change has also had a pronounced effect on biodiversity in general and the diversity of soil organisms in particular. Below-ground diversity is strongly linked to plant biodiversity (Wardle et al., 2004), and often rare plant species or special ecological habitats such as heathlands, wetlands, and Pannonian dry grasslands may harbor unique soil communities. Conversion of such habitats to energy crops would mean losses of above-ground diversity and also rare (and unknown) soil organisms.

Besides land use change, management practices seriously impact soil organisms. Infertile, unproductive ecosystems have soil–food webs characterized by fungal-based energy channels and high densities of enchytraeid worms and macro- and micro-arthropods that may not be found in more fertile productive ecosystems. Any intensification of management such as fertilization or irrigation affects biodiversity in general and soil communities in particular (Figure 3.5).

BIOFUEL PRODUCTION FROM ANNUAL PLANTS

The production of crops for biofuels such as cereals and oilseed rape is likely to cause limited changes in the condition and status of agricultural soils as long as production methods remain unchanged. However, if the management changes, for example, if crop residues are continuously removed to produce cellulosic ethanol, a long-term decline of SOM and negative consequences for soil organisms will result. If no-tillage methods are introduced to conserve soil organic matter, negative impacts on soil organisms may be triggered by enhanced use of pesticides.

The European Environment Agency (EEA 2007) suggests certain energy crops for different climatic zones and pinpoints the risks of planting certain crops in specified regions. High risks of nitrogen losses from soils are predicted for oilseed rape, sugar beet, and linseed in temperate and northern regions; maize and potatoes pose risks in the Mediterranean. For the year 2030 a shift from traditional arable biofuel crops (rape seeds, sunflower seeds, sugar beets, maize corn, wheat–corn, barley–triticale–corn) toward whole-plant cropping of maize, triticale, wheat, sweet sorghum (Italy and Spain), and two-culture options is planned. Sorghum has the best yields of all these annuals, but presents high potential for erosion on sloping soils and subsequent negative consequences for soil organisms (Hallam et al., 2001).

In spite of all the possible negative consequences for the ecology of soils, certain cropping practices are recommended to prevent soil degradation. The following management practices may be beneficial for soil fauna and microflora:

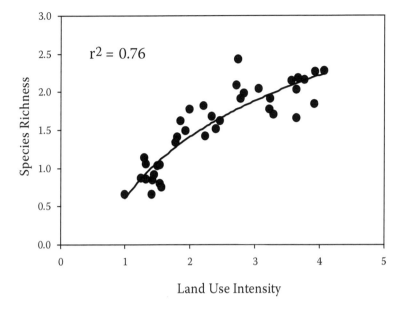

FIGURE 3.5 Biodiversity (species richness) of ants, phanerogames, carabid beetles, mosses, butterflies, snails, spiders, birds) as a function of declining land use intensity (hemerobic state) in agricultural landscale. 1 = highest land use intensity. 4 = lowest land use intensity. (*Source:* Created from data of Zechmeister, H.G. et al., 2003. In Matouch, S. and N. Sauberer, Eds., *International Conference on Predicting Biodiversity in European Landscapes: Mapping, Patterns, Indicators, Monitoring.* Vienna, Nov. 2001. Austrian Federal Ministry for Education, Science and Culture.)

- Introduction of a mix of biomass crops, innovative low input–high yield farming practices, such as mulch systems, double cropping, mixed cropping (winter rye with winter peas; maize with sunflower; leindotter with peas), and strip cropping (Weik et al., 2002).
- Almost all-year soil coverage to promote SOM content and water holding capacity and reduce soil erosion.
- Planting of drought-resistant high yield crops in arid zones.
- Development of individual solutions for biomass cropping while enhancing farm land biodiversity, reducing land use intensity, and preventing soil erosion and fire risk.
- Prevention of conversion of low intensity farm uses such as grasslands, extensive olive cultivation, and other production to biomass cropping; prevention of drainage and plowing.
- Renunciation of rotation of more intensive crops; decreases in irrigation, pesticide and fertilizer inputs, and mechanization.
- Avoidance of landscape changes such as removal of hedges and field boundaries that may act as important retreats for certain development stages of soil organisms such as beetles and diptera.

BIOFUEL PRODUCTION FROM PERENNIAL PLANTS

Among the many perennial plant species used for biofuel production, most lingo-cellulosic crops are grown for the production of second-generation fuels. Positive effects on soil properties, biodiversity, energy balance, GHG mitigation, and carbon footprint are likely when growth is compared to annual arable crops. When compared to steps such as replacement of set-aside and permanently unimproved grassland, such benefits are less apparent (Rowe et al., 2008). However, a large variety of plants are presently under discussion; all have special properties and exert varying effects on soil organisms and soil functions.

INTENSIVE HARVESTING OF OLD-GROWTH FORESTS

In recent decades, forest area and productivity in Europe and North America have increased because of reforestation of marginal agricultural land, recovery from over-utilization of biomass (e.g., litter raking), and some effects of global change such as temperature rises, CO_2 concentrations, and enhanced atmospheric N input. One key facet of the European Union Forest Action Plan (CEC, 2006) is the promotion of the use of this additional forest biomass for energy generation (Katzensteiner and Englisch, 2007).

While timber contains low amounts of nutrients, especially nitrogen, the removal of other plant parts such as branches, needles, leaves, and roots may lead to nutrient depletion and soil acidification. According to Katzensteiner (2008), this may be tolerated for final harvests because it promotes the regrowth of ground vegetation that can reduce leaching losses. In selection systems or in thinning operations, however, it is more crucial to leave residues on site to promote the growth of remaining trees. In addition, this remaining dead wood has a prominent role in soil biodiversity: It offers refuge and breeding space for a large variety of harvestmen, spiders, wood-lice, centipedes, and millipedes (Jabin et al., 2004).

Severe damage to soil and its organisms must be expected after stump removal. Erosion, the disturbance of soil structure, and high leaching rates are consequences of such operations. Soil compaction must be minimized during harvesting operations to maintain the water cycle and soil gas exchange. Different sites present varying vulnerabilities and a particularly sensible system will utilize shallow soils like leptosols, where an organic soil layer as the main rooting zone must be maintained (Katzensteiner, 2008). Also, careful partial harvesting for biofuels in open woodlands and retention of large oaks may present a case of a biodiversity-friendly bioenergy-providing forest system (Götmark, 2007). It is important that case-to-case evaluations of the suitability of a forest stand for biofuel harvesting are made and protocols detailing environmental consequences of intensive harvesting such as those suggested by Smith (1995) for the International Energy Agency Bioenergy Agreement are developed.

SHORT ROTATION COPPICE (SRC) PLANTATIONS

If located, designed, and managed wisely, energy crop plantations of willow, poplar, and oak can produce renewable energy and also generate local environmental benefits. Willow plantations yield soil carbon accumulation, increased soil fertility, reduced nutrient leaching and erosion, and removal of cadmium, crude oil and PAHs from the soil (EEA, 2007; Vervaeke et al., 2003). Metal uptake may be enhanced by siderophores produced by bacteria living in the willow rhizosphere (Kuffner et al., 2008). Another benefit is using willow plantations as vegetation filters to treat nutrient-rich, polluted waters such as municipal wastewater and drainage. However, Jug et al. (1999) demonstrated that planting of SRC on former grasslands led to a loss of nitrogen caused by increased microbial activity and nitrogen mineralization at a rate in excess of plantation requirements.

A special case of SRC is eucalyptus plantation. Although extremely productive, this tree species is known to exert negative effects on soil porosity, making soil a less valuable habitat for certain fauna. In addition, soil organic carbon, total soil nitrogen, and the sizes of soil microbial and microfungal biomasses may decrease (Behera and Sahani, 2003). Polyphenol and volatile terpene present in eucalyptus leaves are known to inhibit microbial enzyme action during decomposition, which may lead to adverse effects on soil fertility.

In Germany the diversity and occurrence of soil microorganisms and soil fauna, especially decomposers such as earthworms, wood lice, harvestmen, and carabides were generally greater in energy forest cultivations than in annual food crop cultivations (Makeschin, 1994). This is mainly due to reduced soil tillage and use of agrochemicals and increased input of litter. The average dose of chemical pesticides applied in willow cultivations is, for example, about 0.2 kg active ingredients ha^{-1} yr^{-1} and about 1 kg in annual food crop cultivation (Börjesson, 1999). If the management of SRC is more intensive, the implied positive effects on soil organisms decline. Sage (1998) argues that tolerating the presence of certain insects and other plants is the main reason why SRC, if appropriately located, has the potential to increase biodiversity in many farm situations and thus takes the first steps in devising ecological guidelines for the establishment of SRC plantations.

In addition to willows in wide use for short rotation energy forestry, other broadleaf tree species like alder, birch, poplar, and hybrid aspen can be used similarly. Among these, alder is special in the sense that it is nitrogen-fixing and thus less dependent on nitrogen fertilization (Perttu, 1995). Small habitats such as cairns, open ditches, and forest edges are very important for biodiversity. They serve as retreats for animals and plants. Such islands should be preserved and must not be shaded by nearby energy forest stands. Certain areas such as old meadows, pastures, paddocks, and wetlands that provide great ecological value as special habitats for fauna should not be used for energy forestry (Perttu, 1995).

Despite the benefits of willow vegetation filters, several potential barriers prevent their large-scale implementation. The first is a lack of knowledge about the risk of spreading pathogens. Large-scale monocultures should be avoided. Row, strip, and alley cropping can increase landscape diversity, thus enhancing biodiversity, helping to prevent wind erosion, and decreasing nitrate leaching to surface waters. A specific application of this system for the Mediterranean would be creating strips of holm oak (EEA, 2007).

Use of Grasses for Bioethanol Production

Certain grass species have been designated for large-scale bioethanol production in Europe: miscanthus (*Miscanthus* spp.), switchgrass (*Panicum virgatum* L.), reed canary grass (*Phalaris arundinacea* L.), and giant reed (*Arundo donax* L.). The first two exhibit C_4 metabolism—they use sunlight more efficiently than the other two species that utilize C_3 metabolism. Another energy efficient C_4 grass is sugar cane (*Saccharum officinarum*), one of the most prominent energy crops of the tropics. Little is known about the consequences of large-scale deployment of the miscanthus compared to SRC willow and poplar, including effects on biodiversity and hydrology (Rowe, 2008).

A study of four miscanthus plantations in Germany reported significant increases in SOM storage in topsoil compared to the grassland control over four years (Kahle et al., 2001). The increases were attributed to the extensive root systems resulting in large below-ground biomass storage. The SOM in miscanthus plots was enriched with lipids, sterols, and free fatty acids and lower in nitrogen-containing compounds, leading to increased hydrophobicity. The authors see a potential for improved physical soil properties due to the role these compounds play in soil aggregate formation and stability. However hydrophobicity may imply adverse conditions for many soil organisms living in the water films of soil pores, e.g., many bacteria and their grazers, like amoebae, ciliates, and nematodes that are essential for the recycling of nutrients. It is likely that commercial miscanthus crops will be fertilized with recommended yearly applications of 88 kg N ha^{-1}, which may result in relatively high nitrate leaching in the first three years of establishment (Christian and Riche, 1998).

Comparing diversity effects of reed canary grass plantations with miscanthus, it seems that invertebrates are at advantage in miscanthus plantations. However, it was suggested that as miscanthus matures, diversity may decline within the cropped area (Semere and Slater, 2005).

It is generally expected that miscanthus and SRC will demand more water than arable crops due to a combination of higher growth rates, high transpiration rates, longer seasonal growth, and increased rooting depth and complexity (Stephens et al., 2001). While demand may lead to irrigation in drier climates, the combined effect of soil drying, decreased runoff, and increased penetration may help reduce flooding in risk areas. Keys to success with these crops include the careful siting of plantations, effective management plans, and development of efficient processing methods, particularly for liquid biofuels.

A comparison of miscanthus, switchgrass, reed canary grass, and giant reed reveals that reed canary grass is the only species native to Europe and adapted to cold and moist conditions. This implies that soil organisms associated with this species will be present on planted wet sites. Reed canary grass also has the potential to become a weed but provides food for insects—a feature that only switchgrass shares (EEA, 2007).

An alternative to monoculture systems for biofuel production may be the use of cuttings from permanent, semi-natural grasslands. In Europe, many of these grasslands are presently abandoned because they are not economically attractive for modern farming systems. The challenge is to design biomass production and processing chains that fit the requirements of maintaining high nature-value grasslands. The conditions for efficient yet nature-friendly use of grassland biomass are not easily met. They may be most easily matched on large-scale and productive grassland sites. However, many examples of species-rich and fairly productive grassland complexes are going or have gone out of production for economic or social reasons including EU Natura 2000 sites. For example, large-scale structural and economic change will affect dairy production in the Alps and lead to surplus grass biomass unable to be used in traditional agricultural production (Pötsch et al., 2006).

Based on the strong linkage of above-ground and below-ground biodiversity (Wardle, 2004), the use of species-rich grasslands for bioenergy would be more beneficial than monocultures for maintaining the diversity of soil organisms. Tilman et al. (2006) showed that high-diversity grasslands in North America produced bioenergy yields that were 238% greater than monoculture yields after a decade. Mixed grasslands also have the advantage over switchgrass; for example, they can be grown on more infertile soils that often harbor special communities of soil organisms (Wardle, 2004).

Ethanol from sugar cane-powered cars in Brazil on and off since the 1930s and Brazil's sugar cane ethanol industry is the world's best and expected to get better, says Emma Marris in her 2006 article titled "Drink the best and drive the rest." Sugar cane ethanol is also the world's cheapest to produce, and a doubling of production is expected by 2014.

In spite of the comparably good GHG emission index (Figure 3.4) serious environmental problems are associated with Brazil's sugar cane ethanol industry. Environmental degradation from soil erosion in sugar cane fields is widespread (Sparovek and Schnug, 2001)—around 30 t ha^{-1} yr^{-1}. When soil is lost to erosion, soil organisms are also lost. Smoke pollution from the burning of sugar cane fields before manual harvesting twice a year is another problem leading to further acidification of already poor tropical soils (Krusche et al., 2003). The alternative to burning is mechanized harvesting that may lead to soil compaction. Although sugar cane needs a dry season and does not directly replace rainforests (with their unique flora and fauna), sugar cane will be pushed to displace other activities such as cotton and soya agriculture and livestock cultivation in the Amazon.

OTHER TROPICAL BIOFUEL CROPS (OIL PALM AND *JATROPHA CURCAS*)

For some time, palm oil was seen as an ideal biofuel that allowed cheap biodiesel production. However, many plantations in Indonesia and Malaysia, where 85% of commercial palm oil is grown, were planted on cleared rainforest and threaten precious ecological habitats. Recently, calculations of the carbon count of producing palm oil fuel revealed that the balance is increasingly negative. A report by a Netherlands-based research group claimed some plantations produced far more carbon dioxide than they saved. Because the plantations are seeded on drained peat swamps, they unleash fluxes of carbon from a valuable storage pool that was locked in peat for several million years (Hooijer et al., 2006). In addition, fires on the drained peatlands lead to further soil destruction and carbon loss. Consequently, many companies have withdrawn their plans to expand palm oil biofuel production.

Jatropha curcas is another candidate for producing high quality biofuel for vehicles. The Indian government is stimulating the rapid expansion of Jatropha monocultures for biodiesel on 50 million ha classified as wastelands. Jatropha is widely promoted as a crop that can grow in dry regions, but regular and sufficient rainfall is needed to sustain high yields. In arid and semi-arid areas, fertilizers and irrigation are needed for the first three years of cultivation. In large parts of India, groundwater tables are falling, threatening the futures of large agriculture areas. It is feared that the promotion of Jatropha for bioenergy may destroy primary and secondary forests in India, seriously harming above- and below-ground biodiversity (Ernsting et al., 2008). However, if used carefully on degraded land in the form of agroforestry systems, Jatropha has the potential to provide more environmentally friendly biodiesel production.

BIOFUELS AS SOIL POLLUTANTS

Ethanol has chemical properties unlike those of traditional gasoline. It can easily lead to corrosion of auto engine components, fuel tanks, service station equipment, and pipelines, increasing the risk of leakage into surrounding soils. Pipelines for ethanol have long been considered problematic because ethanol tends to absorb moisture and impurities as it flows.

Ethanol reduces the interfacial tension of gasoline, allowing the ethanol–gasoline non-aqueous phase liquid to enter smaller pore spaces in the soil and infiltrate more easily through the vadose zone to the water table. Alcohols are known to cause the dehydration of both swelling and non-swelling clays, producing microfractures that increase clay permeability (Niven, 2005). We should remember that increased use of bioethanol presents pollution implications for soil organisms and poses new problems for soil remediation.

OUTLOOK

The effect of biofuel production on soil organisms depends on (1) the siting of the biofuel production system and whether it involves land use change; (2) the nature of the biofuel crop, whereby mixed grasslands and diverse forests are more beneficial to soil organisms than monocultures of any kind; and (3) land use intensity and management techniques, some of which can reduce adverse effects on soil biota and promote biodiversity. The importance of soil protection is illustrated by the current search for biotechnological techniques to increase the efficiency of biofuel production, where organisms and enzymes extracted from soils play a major role. It has to be kept in mind that soils are a non-renewable resource and their degradation may lead to an irreversible loss of gene reserves for the future.

A relief of the pressure on soils for the production of bioenergy plants might come from a third generation of biofuels, made from microalgae (Huntley and Redalje, 2007). According to the authors these can produce up to 50,000 liters of lipid per hectare per year compared with only 2,400 liters obtained from high yield oil plants. In addition, the costs have been estimated at $50 or less per barrel, making it more economic at current oil prices without subsidies. The technology of aquaculture seems to be well-developed, and as land-based algal culture systems don't depend on soil fertility, barren land could be used, saving soils for food production and the sustenance of natural habitats.

POSSIBLE ECOLOGICAL AND ECONOMIC IMPACTS OF AGRICULTURAL BIOMASS PRODUCTION ON GAME MANAGEMENT

Game management is intimately connected with human land use. The biofuel option creates additional pressure to use marginal lands and intensify agricultural production on already used areas. Both factors will exert significant impacts on wildlife and wildlife management. We will discuss the European situation as an example, although we are well aware that conflicts involving intensification of agriculture, nature conservation, and game management are widespread everywhere. The elephant–agriculture conflict in Africa is one of the most debated situations (Kiley-Worthington, 1997; Hoare, 2000).

Europe's agriculture is dominated by EU's Common Agricultural Policy (CAP). From the beginning, the policy aimed to supply the EU with staple foods and assure productivity and food security. Initially, this led to the famous "milk lakes" and "butter mountains" that were reduced over the years as farmer subsidies targeted food quality and the management of natural habitats. One of the CAP results was the abandonment of arable land, decreasing potential acreage for cereal production about 10%. Farmers had the option to set aside the land or produce renewable resources on it. Since the latter was not economically beneficial, most of the set-aside land was not used. In principle, this policy provided a more diverse landscape offering protection and food for many game species such as hares and partridges found on Europe's arable land. As a result, hunters and conservationists benefited from the CAP.

The cultivation of bioenergy plants on set-aside properties became more attractive for farmers as a result of national and EU funding. At the same time, prices for food greatly increased. Hence, each hectare will face constant use in the competition between food and biofuel production. The 10% rule was neglected due to the shortage of food, demands for food exports, and the need for biofuel. For small game species, this meant significant losses of habitat or habitat quality.

Fallow land with low vegetation has been displaced by intensively crops that produce maximum vegetation; such land does not represent attractive habitats for species originating from steppe environments. Tall cornfields limit the abilities of small mammals and ground nesting birds to inspect their surroundings. The soil dries more slowly after rainfall; hares do not find preferred herbs and partridges cannot find insects for their young. Conversely, control of predators like foxes, badgers, and raccoons is not feasible in an area of perfect cover like a cornfield. Moreover, energy crops are harvested at a time when biomass is the largest and materials the freshest. This means renewable resources are harvested during game species reproduction periods.

Modern harvesters work at high speed and at night to increase efficiency and contain costs. It is likely that more pastures will be used for the production of renewable resources and the result will be loss of valuable habitats for numerous songbird species. In sum, these factors will keep population growth rates in game species low.

Within a short time, populations of hares, pheasants, and partridges will decrease further along with other species typically found in open landscapes with low vegetation (Bro et al., 2004; Smith et al., 2005; Jennings et al., 2006; Watson et al., 2007).

Another consequence of enhanced biomass growth will be an increase in crop damage caused by game animals. While the situation for small game appears grim, other animals will benefit from the high cover structures of cornfields. For example, wild boar will inhabit cornfields for a few months out of the sight of hunters and will cause damage to the cornfield and adjacent land (Schley et al., in press). Rising prices for agricultural products (food and biomass) mean increases in game damage compensation payments, deteriorating the already difficult situation of game damage caused by wild boar. This raises the question whether hunters will be interested in highly arable areas in the future if the situation is characterized by game damage and compensation costs. Without hunters, small villages in rural areas will lose income from hunting rents and incur high costs for damage compensation.

This situation should yield a new strategy of game management for wild boar. Recent studies show that game damage positively correlates with game abundance (Geisser and Reyer, 2004; Herrero et al., 2006). Food supplies for wild boar may reduce damage, but only if the boar density is below 1.5 individuals 100 ha^{-1}. Additional food supplies produced no effects or led to further proliferation of population numbers (Calenge et al., 2004). A more effective method than feeding to reduce wild boar damage is a sustainable harvest strategy with flexible harvest rates depending on climate conditions and food availability (Bieber and Ruf, 2005). Hunters have to leave the stand at the feeding place and develop approaches for active hunts with their neighbors.

To counteract the gloomy forecasts for the development of hunting in times of biofuel production, hunters should attempt to influence the future of biofuel production by shedding light on the impacts of biofuel production on landscapes. We lose more than the regional character of our cultural landscapes. We risk further declines in biodiversity by limiting biocultural diversity.

CONCLUSIONS

Removal of crop residues, for example, to produce cellulosic ethanol, can entail a long-term decline of SOM. This is obviously an unsustainable practice. It compromises soil functional integrity and may trigger adverse effects such as increased runoff and erosion. These impacts are strongly dependent on crop types, soil properties, and climatic conditions of a specific location.

With continued use, soils become more compacted and do not have time to recover. As a result they will not be sustainable for long-term agricultural growth for bioenergy or fuel production purposes. High levels of compaction severely inhibit plant growth and negatively affect biomass productivity, GHG emissions, and sustainability. To reduce the negative and additive compaction effects, permanent harvesting lanes must be established. GHG balances of biofuel production systems are heavily influenced by soil deterioration, nitrogen use efficiency, and by-products for additional uses such as feedstuffs.

Future biofuels are expected to be produced from cellulose and lignin; this means most parts of the plants grown for fuel production will be utilized. Therefore the area-based yields of second generation biofuels are expected to be considerably higher than yields from current production. The target compounds contain little nitrogen, and it is expected that N fertilizer rates can be kept low. However, because fuel crops cannot be produced on current agricultural sites best suited for farming and critical to filling future food needs, it remains to be seen whether optimizing yields would lead, as in classical agriculture, to over-fertilization to an extent that will offset expected improvements in the GHG balances of biofuels of the second generation.

Some relief of the pressure on soils from an ecological view may come from a third generation of biofuels—microalgae that can produce up to 50,000 liters lipid $ha^{-1} yr^{-1}$ compared with only 2,400 liters from high yield oil plants.

The effects of biofuel production on soil organisms depend on (1) the siting of the biofuel production system and whether it involves land use change, (2) the nature of the biofuel crop; mixed grasslands and diverse forests are more beneficial to soil organisms than any monoculture, and (3) land use intensity and management techniques, some of which can reduce adverse effects on soil biota and promote biodiversity. Considerable problems may arise from impacts on game animals. Wild boar may continue to spread and further proliferation of populations are likely. Damage to crops, inhibition of effective hunting by sheltering wild boar, and increases of compensation payments must be considered.

REFERENCES

Alvarez, R. 2005. A review of nitrogen fertilizer and conservation tillage effects on soil organic carbon storage. *Soil Use Mgt.* 21: 38–52.

Antil, R.S. et al. 2005. Long-term effects of cropped versus fallow and fertilizer amendments on soil organic matter I: Organic carbon. *J. Plant Nutr. Soil Sci.* 68: 108–116.

Arshad, M.A. and Martin, S. 2002. Identifying critical limits for soil quality indicators in agro-ecosystems. *Agric. Ecosyst. Environ.* 88: 153–160.

Aufhammer, W. 1998. *Getreide- und andere Körnerfruchtarten.* Ulmer, Stuttgart.

Bai, Z.G. et al. 2008. Proxy global assessment of land degradation. *Soil Use Mgt.* 24: 223–234.

Behera, N. and Sahani, U. 2003. Soil microbial biomass and activity in response to eucalyptus plantation and natural regeneration on tropical soil. *Forest Ecol. Mgt.* 174: 1–11.

Bieber, C. and Ruf, T. 2005. Population dynamics in wild boar *Sus scrofa*: Ecology, elasticity of growth rate and implications for the management of pulsed resource consumers. *J. Appl. Ecol.* 42: 1203–1213.

Birkas, M. 2008. *Environmentally Sound Adaptable Tillage.* Akademiai Kiado, Budapest.

Blum, W.E.H. 2005. Functions of soil for society and the environment. *Rev. Environ. Sci. Biotechnol.* 4: 75–79.

Blum, W.E.H. and Eswaran H. 2004. Soils for sustaining global food production. *J. Food Sci.* 69: 37–42.

Börjesson, P. 1999. Environmental effects of energy crop cultivation in Sweden I: Identification and quantification. *Biom. Bioenergy* 16: 137–154.

Bouwman, A.F. et al. 2002. Emissions of N_2O and NO from fertilized fields: Summary of available measurement data. *Global Biogeochem. Cycles* 164: 1058.

Bro, E. et al. 2004. Impact of habitat management on grey partridge populations: Assessing wildlife cover using a multisite BACI experiment. *J. Appl. Ecol.* 41: 846–857.

Calenge, C. et al. 2004. Efficiency of spreading maize in the garrigues to reduce wild boar *Sus scrofa* damage to Mediterranean vineyards. *Eur. J. Wildlife Res.* 50: 112–120.

Carter, M.R. 2002. Soil quality for sustainable land management: Organic matter and aggregation interactions that maintain soil functions. *Agron. J.* 94: 38–47.

CEC 2006 EU Forest Action Plan. http://www.cec.org/files/PDF/PUBLICATIONS/EDrpt-Council2006_en.pdf 02.06.2006.

Christian, D.G. and Riche, A.B. 1998. Nitrate leaching losses under miscanthus grass planted on a silty clay loam soil. *Soil Use Mgt.* 14: 131–135.

Couteaux, M.M. et al. 2001 Decomposition of C-13-labelled standard plant material in a latitudinal transect of European coniferous forests: Differential impact of climate on the decomposition of soil organic matter compartments. *Biogeochemistry* 54: 147–170.

Crutzen, P.J. 2002. Geology of mankind. *Nature* 415, 23.

Crutzen P.J. et al. 2008. N_2O release from agro-biofuel production negates climate effect of fossil fuel derived CO_2 savings. *Atm. Chem. Phys.* 8: 389–395.

Davidson, E.A. and Janssens, I.A. 2006. Temperature sensitivity of soil carbon: Decomposition and feedbacks to climate change. *Nature* 440: 165–173.

EEA. 2007. Estimating the environmentally compatible bioenergy potential from agriculture. European Environment Agency Report 12/2007. http://reports.eea.europa.eu/technical_report_2007_12/en/Estimating_the_environmentally_compatible_bio-energy_potential_from_agriculture.pdf.

EEA. 2006. How much bio energy can Europe produce without harming the environment? European Environment Agency Report 7/2006.

Emmett, B.A. et al. 2004. The response of soil processes to climate change: Results from manipulation studies of shrub lands across an environmental gradient. *Ecosystems* 7: 625–637.

Ernsting, A. et al. 2007. Agrofuels: Toward a reality check in nine key areas. Twelfth Meeting of the Subsidiary Body on Scientific, Technical and Technological Advice of the Convention on Biological Diversity. Paris, July 2007. http://www.biofuelwatch.org.uk/docs/agrofuels_reality_check.pdf 06/2007.

Fallon, P. et al. 1998. Estimating the size of the inert organic matter pool for use in the Rothamsted carbon model. *Soil Biol. Biochem.* 30: 1207–1211.

Fazekas, O. 2005. Bedeutung von Bodenstruktur und Wasserspannung als stabilisierende Kenngrößen gegen intensive mechanische Belastungen in einer Parabraunerde aus Löss unter Pflug- und Mulchsaat Schriftenreihe des Instituts für Pflanzenernährung und Bodenkunde. CAU Kiel, H.67.

Fazekas, O. and Horn, R. 2005. Interaction between mechanical and hydraulically affected soil strength depending on time of loading. *Z. Pflanz. Boden.* 168: 60–67.

Galloway, J.N. et al. 2004. Nitrogen cycles: Past, present, and future. *Biogeochemistry* 70: 153–226.

Galloway, J.N. et al. 2008. Transformation of the nitrogen cycle: Recent trends, questions and potential solutions. *Science* 320: 889–892.

Geisser, H. and Reyer, H.U. 2004. Efficacy of hunting, feeding, and fencing to reduce crop damage by wild boars. *J. Wild. Mgt.* 68: 939–946.

Gerzabek, M.H. 2007. Soil organic matter dynamics under various land use and management. *Ann. Agrar. Sci.* 5: 18–22.

Gerzabek, M.H. et al. 2005. Quantification of organic carbon pools for Austria's agricultural soils using a soil information system. *Can. J. Soil Sci.* 85: 491–498.

Gerzabek, M.H. et al. 1997. The response of soil organic matter to manure amendments in a long-term experiment in Ultuna, Sweden. *Eur. J. Soil Sci.* 48: 273–282.

Götmark, F. 2007. Careful partial harvesting in conservation stands and retention of large oaks favour oak regeneration. *Biol. Conserv.* 140: 349–358.

Guo, L.B. and Gifford, R.M. 2002. Soil carbon stocks and land use change: A meta analysis. *Global Change Biol.* 8: 345–360.

Haberl, H. et al. 2007. Quantifying and mapping the human appropriation of net primary production in earth's terrestrial ecosystems. *Proc. Natl. Acad. Sci. USA* 104: 12942–12947.

Hallam, A., Anderson, I.C., and Buxton, D.R. 2001. Comparative economic analysis of perennial, annual, and intercrops for biomass production. *Biom. Bioenergy* 21: 407–424.

Herrero, J. et al. 2006. Diet of wild boar *Sus scrofa* L. and crop damage in an intensive agroecosystem. *Eur. J. Wild. Res.* 52: 245–250.

Hoare, R. 2000. African elephants and humans in conflict: Outlook for co-existence. *Oryx* 34: 34–38.

Hooijer, A. et al. 2006. PEAT-CO_2: Assessment of CO_2 emissions from drained peatlands in SE Asia. *Delft Hydraul. Q. Rep.* 3943: 41. http://www.wetlands.org/LinkClick.aspx?fileticket=NYQUDJl5zt8%3d&language=en-US 07.12.2006.

Horn, R. 1988. Compressibility of arable land. *Catena Suppl.* 11: 53–71.

Horn, R. 2004. Time dependence of soil mechanical properties and pore functions for arable soils. *Soil Sci. Soc. Am. J.* 68: 1131–1137.

Horn, R. et al. 2002. Prediction of soil strength of arable soils and stress dependent changes in ecological properties based on maps. *Z. Pflanz. Boden.* 165: 235–240.

Horn, R. et al. 2006. *Soil Management for Sustainability*. Advances in Geoecology Series 38, Catena Publishing.

Horn, R. et al., Eds. 2000. *Subsoil Compaction: Distribution, Processes and Consequences*. Advances in Geoecology Series 32, Catena Publishing.

Huntley, M.E. and Redalje, D.G. 2007. CO_2 mitigation and renewable oil from photosynthetic microbes: New appraisal. *Mitigation Adapt. Strat. Global Change* 12: 573–608.

IPCC. 2006. *Guidelines for National Greenhouse Gas Inventories* prepared by the National Greenhouse Gas Inventories Programme, Vol. 4, Eggleston, H.S. et al., Eds. IGES, Hayama, Japan, Chap. 4.

Jabin, M. et al. 2004. Influence of deadwood on density of soil macro-arthropods in a managed oak–beech forest. *Forest Ecol. Mgt.* 194: 61–69.

Jenkinson, D.S. et al. 1992. Calculating net primary production and annual input of organic matter to soil from the amount and radiocarbon content of soil organic matter. *Soil Biol. Biochem.* 24: 295–308.

Jenkinson, D.S. and Coleman, K. 1994. Calculating the annual input of organic matter to soil from measurements of total organic carbon and radiocarbon. *Eur. J. Soil Sci.* 45: 167–174.

Jug, A. et al. 1999. Short-rotation plantations of balsam poplars, aspen and willows on former arable land in the Federal Republic of Germany III: Soil ecological effects. *Forest Ecol. Mgt.* 121: 85–99.

Kahle, P. et al. 2001. Cropping of miscanthus in Central Europe: Biomass production and influence on nutrients and soil organic matter. *Eur. J. Agron.* 15: 171–184.

Kaiser, K., Mikutta, R., and Guggenberger, G. 2007. Increased stability of organic matter sorbed to ferrihydrite and goethite on aging. *Soil Sci. Soc. Am. J.* 71: 711–719.

Katzensteiner, K. 2008 Forest biomass and bioenergy: Considerations for sustainability and ecosystem services. *Local Land Soil News* 24/25: 19–20.

Katzensteiner, K. and Englisch, M. 2007. Sustainable biomass production from forests: Lessons from historical experience and challenges for ecological research. *Austr. J. Forest Sci.* 124: 201–214.

Keller, T. 2004. *Soil Compaction and Soil Tillage: Studies in Agricultural Soil Mechanics*. Swedish University of Agricultural Sciences.

Kiley-Worthington, M. 1997. Wildlife conservation, food production and development: Can they be integrated? *Environ. Values* 6: 455–470.

Kirchmann, H. et al. 2004. Effects of level and quality of organic matter input on soil carbon storage and biological activity in soil. *Global Biogeochem. Cycles* 18: 1–9.

Krausmann, F. et al. 2008. Global patterns of socioeconomic biomass flows in the year 2000: Comprehensive assessment of supply, consumption and constraints. *Ecol. Econ.* 65: 471–487.

Krusche, A.V. et al. 2003. Acid rain and nitrogen deposition in a sub-tropical watershed Piracicaba: Ecosystem consequences. *Envir. Poll.* 121: 389–399.

Kuffner, M. et al. 2008. Rhizosphere bacteria affect growth and metal uptake of heavy metal accumulating willows. *Plant Soil*, in press.

Kühner, S. 1997. Simultane Messungen von Spannungen und Bodenbewegungen bei statistischer und dynamischer Belastung zur Abschätzung der dadurch induzierten Bodenbeanspruchung; Schriftenreihe Institut für Pflanzenernährung und Bodenkunde der Universität Kiel.

Lal, R. 2006. Soil science in the era of hydrogen economy and 10 billion people. In *The Future of Soil Science*, IUSS, pp. 76–79.

Lynd, L.R. et al. 2002. Microbial cellulose utilization: Fundamentals and biotechnology. *Microbiol. Mol. Biol. Rev.* 66: 506–577.

Makeschin, F. 1994. Effects of energy forestry on soils. *Biom. Bioenergy* 6: 63–80.

Malhi, S.S. et al. 2006. Tillage, nitrogen and crop residue effects on crop yield, nutrient uptake, soil quality, and greenhouse gas emission. *Soil Tillage Res.* 90: 171–183.

McKendry, P. 2002. Energy production from biomass, Part 1: Overview of biomass. *Biores. Technol.* 83: 37–46.

Niven, R.K. 2005. Ethanol in gasoline: Environmental impacts and sustainability review article. *Renew. Sustain. Energy Rev.* 9: 535–555.

Oberländer, H.E. and Roth, K. 1974. Ein Kleinfeldversuch über den Abbau und die Humifizierung von ^{14}C-markiertem Stroh und Stallmist. *Bodenkultur* 25: 111–129.

Pacala, S. and Socolow, R. 2004. Stabilization wedges: Solving the climate problem for the next 50 years with current technologies. *Science* 305: 968–972.

Pagliai, M. and Jones, R. 2002. Sustainable land management and environmental protection: A soil physical approach. *Adv. Geoecol.* 35: 588.

Peng, X. and Horn, R. 2008. Anisotropic responses of soil strength and pore functions to 10-year previous tractor wheeling under conservation and conventional tillage. *J. Plant Nutr. Soil Sci.*, in press.

Perttu, K.L. 1995. Ecological, biological balances and conservation. *Biom. Bioenergy* 9: 107–116.

Peth, S. et al. 2006. Heavy soil loading and it consequences for soil structure, strength and deformation of arable soils. *J. Plant Nutr. Soil Sci.* 169: 775–783.

Peth, S. and Horn, R. 2006. The mechanical behaviour of structured and homogenized soils under repeated loading. *J. Plant Nutr. Soil Sci.* 169: 401–411.

Piepenbrock, A. 2002. Effect of short term willow growth and harvesting on soil physical properties. BSc Thesis, CAU, Kiel, Germany.

Polyakov, V.O. and Lal, R. 2008. Soil organic matter and CO_2 emission as affected by water erosion on field runoff plots. *Geoderma* 143: 216–222.

Pötsch, E.M. et al. 2006. Effect of different management systems on quality parameters of forage from mountainous grassland. *Grassland Sci. Eur.* 11: 484–486.

Powlson, D.S., Riche, A.B., and Shield, I. 2005. Biofuels and other approaches for decreasing fossil fuel emissions from agriculture. *Ann. Appl. Botany* 146: 193–201.

Prather, M. et al. 2001. Atmospheric chemistry and greenhouse gases. In Houghton, J.T. et al., Eds., *Climate Change 2001: The Scientific Basis*. Cambridge University Press, pp. 239–287.

Quideau, S.A. et al. 2000. Soil organic matter processes: Characterization by ^{13}C NMR and ^{14}C measurements. *Forest Ecol. Mgt.* 138: 19–27.

Rowe, R.L., Street, N.R., and Taylor, G. 2008. Identifying potential environmental impacts of large-scale deployment of dedicated bioenergy crops in the UK. *Renew. Sustain. Energy Rev.*, in press.

Sage, R.B. 1998. Short rotation coppice for energy: Toward ecological guidelines. *Biom. Bioenergy* 15: 39–47.

Sanderson, M.A. et al. 2006. Switchgrass as a biofuel feedstock in the USA. *Can. J. Plant Sci.* 86: 1315–1325.

Schäffer, J., Hartmann, R., and Wilpert, K. 2001. Effects of timber harvesting with tracked harvesters on physical soil properties. In Johannson, J., Ed., *Proceedings of Third and Final Meeting of a Concerted Action: Excavators and Backhoe Loaders as Base Machines in Forest Operations.* Pisa, Italy, Sept. 2000, pp. 119–124.

Schley, L. et al. 2009. Patterns of crop damage by wild boar *Sus scrofa* in Luxembourg over a 10-year period. *Eur. J. Wildlife Res.*, in press.

Schrader, S. and Bayer, B. 2000. Abundances of mites Gamasina and Oribatida and biotic activity in arable soil affected by tillage and wheeling. *Braunsch. Natur. Schrift.* 6: 161–181.

Semere, T. and Slater, F. 2005. The effects of energy grass plantations on biodiversity. B/CR/00782/00/00 URN 04/823. DTI; 2005 http://www.berr.gov.uk/files/file15002.pdf.

Semmel, H. and Horn, R. 1995. Spannungen und Spannungsverteilung in Ackerböden aufgrund von Befahrungen- Überlegungen zur Druckfortpflanzung and der mechanischen Belastbarkeit. *Landbauforsch. Völk.* 147: 41–61.

Smil, V. 2001. *Enriching the Earth: Fritz Haber, Carl Bosch and the Transformation of World Food Production.* MIT Press, Cambridge.

Smith, C.T. 1995. Environmental consequences of intensive harvesting. *Biom. Bioenergy* 9: 161–179.

Smith, R.K. et al. 2005. Vegetation quality and habitat selection by European hares *Lepus europaeus* in a pastural landscape. *Acta Theriol.* 50: 391–404.

Sparling, G.P. 1985. Soil organic matter and biological activity. In Vaughan, D. and Malcolm, R.E., Eds., *The Soil Biomass.* Martinus Nijhoff, Dordrecht, p. 223.

Sparovek, G. and Schnug, E. 2001. Temporal erosion-induced soil degradation and yield loss. *Soil Sci. Soc. Am. J.* 65: 1479–1486.

Stephens, W., Hess, T., and Knox, J. 2001. Review of the effects on energy crops on hydrology. MAFF NF0416 http://www.defra.gov.uk/farm/crops/industrial/research/reports/nf0416.pdf.

Tilman, D., Hill, J., and Lehman, C. 2006. Carbon-negative biofuels from low-input high-diversity grassland biomass. *Science* 314: 1598–1600.

U.S. DOE. 2006. Breaking the Biological Barriers to Cellulosic Ethanol: A Joint Research Agenda. DOE/SC-0095. http://genomicsgtl.energy.gov/biofuels/2005workshop/b2blowres63006.pdf 06/2006

van den Akker, J.J.H., Arvidsson, J., and Horn, R., Eds. 1995. Experiences with impact and prevention in the European Community. Report 168.

Vervaeke, P. et al. 2003. Phytoremediation prospects of willow stands on contaminated sediment: A field trial. *Environ. Pollut.* 126: 275–282.

Vossbrink, J. and Horn, R. 2004. Modern forestry vehicles and their impact on soil physical properties. *Eur. J. Forest Res.* 123: 259–267.

Wardle, D.A. et al. 2004. Ecological linkages between aboveground and belowground biota. *Science* 304: 1629–1633.

Watson, M., Aebischer, N.J., and Cresswell, W. 2007. Vigilance and fitness in grey partridges *Perdix perdix*: Effects of group size and foraging-vigilance trade-offs on predation mortality. *J. Animal Ecol.* 76: 211–221.

Weik, L. et al. 2002. Grain yields of perennial grain crops. *J. Agron. Crop Sci.* 188: 342–349.

Wiermann, C. 1998 Auswirkung differenzierter Bodenbearbeitung auf die Bodenstabilität und das Regenerationsvermögen lößbürtiger Ackerstandorte; Schriftenreihe Institut für Pflanzenernährung und Bodenkunde der Universität Kiel.

Wiermann, C. et al. 2000. Stress/strain processes in a structured silty loam Luvisol under different tillage treatments in Germany. *Soil Tillage Res.* 53: 117–128.

Zechmeister H.G. et al. 2003. Land use intensity: An essential environmental correlate of plant species richness in agricultural landscapes. In Matouch, S. and N. Sauberer, Eds., *International Conference on Predicting Biodiversity in European Landscapes: Mapping, Patterns, Indicators, Monitoring.* Vienna, Nov. 2001. Austrian Federal Ministry for Education, Science and Culture.

Zink, A., Fleige, H., and Horn, R. 2008. Stress-induced changes in soil strength and consequences for hydraulic properties under conventional and conservation tillage. *Soil Tillage Res.*, in preparation.

4 Land Use in Production of Raw Materials for Biofuels

Leidivan Almeida Frazão, Karina Cenciani,
Marília Barbosa Chiavegato, Carlos Clemente Cerri,
Brigitte Josefine Feigl, and Carlos Eduardo P. Cerri

CONTENTS

INTRODUCTION

The problems of oil supply in the world market during the 1930s, combined with efforts of the European countries to develop alternative sources of energy, culminated in the search for viable solutions for fossil fuel replacement. In this scenario, research on biodiesel fuels achieved a place of prominence. Studies focused on obtaining derivative compounds with physical and chemical properties closer to the liquid fuels used in combustion engines of the Otto and Diesel cycles. This would allow the production of a mixture to be added to the fossil fuel or the total replacement of fossil fuel with no need to change engines (Suarez and Maneghetti, 2007).

However, after the Second World War ended, biodiesel studies temporarily ceased due to the stabilization of the world market of oil. In the 1970s and 1980s, researchers and governments regained interest in alternative fuels as a result of increasing concern that energy sources would be depleted unless other natural substitutes were developed. Within this historic context, ethanol emerged as an alternative source of energy to oil. Brazil then became the world's largest producer.

Biodiesel was expected to be the main substitute for oil diesel. In fact, it developed into an alternative to oil-derived fuels that can be used in cars and other vehicles with diesel engines. Made from renewable energy sources, biodiesel emits fewer pollutants than fossil diesel. It has virtually the same properties but it can reduce net emissions of carbon dioxide by 78%, smoke emissions by 90%, and eliminate emissions of sulfur oxide (Holland, 2004). It has been adopted by many countries, especially in Europe, where the main producers are Germany, Italy, and France.

The use of oil as the main energy matrix has contributed over the decades to the increase of greenhouse gases (GHGs) in the atmosphere. The main GHGs responsible for global warming are CO_2 (carbon dioxide), CH_4 (methane), and N_2O (nitrous oxide). To prevent or slow increases of GHG concentrations, some mitigation decisions must be made (Adler et al., 2007). Among a number of strategies proposed by Pacala and Socolow (2004) for mitigation of carbon (C) based on the use of new technologies, the use of biofuels is seen as one of the most viable alternatives. Concerns about climate changes are growing and have led to global policies for emission reductions that are necessary for the transition to a new energy matrix that can replace oil.

Biofuels, especially ethanol and biodiesel, are seen by environmentalists and government leaders as the most promising alternatives to reduce fossil fuel dependency and, as a consequence, reduce CO_2 emissions (Farrell et al., 2006; Ragauskas et al., 2006). Furthermore, in some cases, biofuels can enhance local agricultural support and economic development (Goldemberg et al., 2007). Three general principles should guide the viability of policies and practices related to biofuel use: (1) promotion of sustainability with low ecological footprint while maintaining the lowest possible impact on the supply, (2) sustainability of native systems and essential cultures, and (3) establishment of a requirement for a neutral carbon balance for biofuels (Groom et al., 2008). The urgent need for the reduction of GHG emissions to the atmosphere favors the acceptance of biofuels due to their low CO_2 emissions. All biofuels proposed to date, with the exception of ethanol derived from corn, show great potential for reducing air pollution and decreasing CO_2 outputs (Powlson et al., 2005; Farrell et al., 2006).

An accurate assessment of the productive chain of biodiesel should be based on an environmental approach. An important issue that must be considered is the use of fossil fuels that increase GHG emissions to the atmosphere during production of biodiesel. The largest emissions of GHGs arising from agriculture are N_2O, CO_2, and

CH_4 from soil and CO_2 emissions from agricultural machinery (Robertson, 2000; Del Grosso et al., 2001; West and Marland, 2002).

A cultivation system based on bioenergy varies by plant life cycle, productivity, efficiency of energy conversion, demands for nutrients, inputs of carbon in soil, losses of nitrogen (N), and other characteristics resulting from management operations. These factors affect the magnitude of the components that contribute to net fluxes of GHGs and losses of N. N_2O emissions and NO_3 leaching vary with the amount of nitrogen fertilizer applied and its interaction with precipitation, temperature, soil texture, and crop rotation (Adler et al., 2007).

LAND USE AND EXPANSION OF AGRICULTURAL AREA
FOR PRODUCTION OF RAW MATERIALS

Two main points to be considered when discussing the production of biofuels are land use and occupation of areas that could be used for agriculture. In Brazil, some areas now used by the livestock sector, if better managed, could allow the expansion of agricultural cultivation of grains and oil seeds. Livestock farming in Brazil is extensive because of the availability of large areas of land. Preliminary results released in 2006 by the Instituto Brasileiro de Geografia e Estatística (IBGE) show that Brazilian pasture areas total approximately 172 million ha, covering almost 50% of the agricultural area. However, inadequate pasture management caused the degradation of soil, leading to land abandonment and clearing of forest areas for use as new pastures (Demattê, 1998). Estimates of the Instituto Nacional de Pesquisas Espaciais (INPE, 2000) show that deforested areas account for more than 551,000 km^2. Of the total deforested area, 45% is pasture, 28% secondary forests originated from pastures abandoned after 1970, and 2% degraded pasture (Fearnside, 1996).

Pasture areas in the north region of Brazil increased from 24.3 million ha in 1995 to 32.6 million ha in 2006, representing a 33.8% increase of lands used for livestock (IBGE, 2006). There was country-wide shrinkage of 3% in land use for livestock, with decreases in the south, southeast, and west central regions. In the north, the increase of pasture appears tied to the increase of almost 14 million head of cattle representing about 80.7% of the total herd (IBGE, 2008).

Areas for agricultural use in the Amazon region show serious problems regarding the failure to conserve natural resources. Currently, about 60% of the area covered with pastures is in a state of advanced degradation. Degraded pastures (Figure 4.1A) are characterized by lack of soil nutrients, low plant biomass, few seeds of primary forest, weeds in large quantities, lack of forest seeds in seed banks, low rates of germination (Nepstad et al., 1998), and low soil drainage and compaction levels (Eden et al., 1991). The degradation of pastures also results from inadequate management of cattle herds and lack of corrective fertilization and pasture maintenance (Macedo, 2000).

The Brazilian system of livestock production must be restructured. Appropriate management of cattle and pastures (Figure 4.1B) depends on the increasing soil productivity, preventing deforestation, and devising more sustainable systems. The use of degraded pasture areas for growing food crops can reduce the competition for land between food crops and crops with potential to produce biofuels including oil plants.

FIGURE 4.1   Degraded (a) and well-managed (b) pastures in Brazil.

The appropriate management of pastures should include the potential for carbon sequestration which, combined with the use of beef tallow in biodiesel production, can reduce GHG emissions to the atmosphere and help mitigate global warming.

Another interesting aspect to consider relates to the use of lands that have not been used for food crops. Areas in the semi-arid northwest region of Brazil are not suitable for cultivation of grains because they have not adapted to edaphoclimatic conditions of the region. However, those areas can be used for cultivation of oil crops such as castor beans (*Ricinus communis* L.) and *Jatropha curcas* L. that are perfectly suited to such conditions and less demanding of water and nutrients.

RAW MATERIALS FOR BIODIESEL PRODUCTION IN BRAZIL

OILSEEDS

The production of oil from agricultural crops is increasing because of the growing demand for oil used in biodiesel production. Large agricultural areas of Brazil are capable of inclusion in this process, particularly in the Amazon region and some northeast states (Figure 4.2). For oilseed cultivation, it is necessary to know the productive capacity of each species in different regions to allow expansion of cultivation. There is incentive for sustainable extraction of native species and producing perennial legumes using available technology.

In addition to the environmental and economic benefits, agricultural production intended for generating energy in countries like Brazil can serve as a viable alternative for small farmers. Focusing on oilseed cultivation, it is possible to create many other opportunities along the production chain that will generate jobs and income. The following is a brief summary of various oilseeds with potential to produce biodiesel in Brazil. The oilseeds receiving attention from the National Program for Production and Use of Biodiesel (PNPB) are castor beans and palm (*Elaeis guineensis* Jacq.)

Soybean

The soybean (*Glycine max* L.; Figure 4.3) is considered a major source of protein and vegetable oil worldwide. The seeds originated in temperate climates, but sequential breeding programs made it possible to expand cultivation to subtropical and

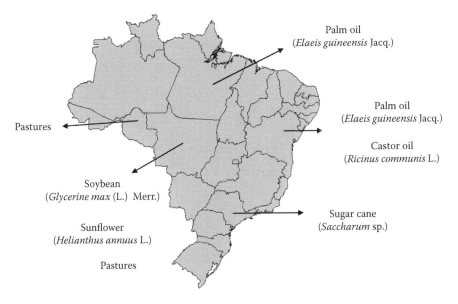

Palm oil
(*Elaeis guineensis* Jacq.)

Palm oil
(*Elaeis guineensis* Jacq.)

Pastures

Castor oil
(*Ricinus communis* L.)

Soybean
(*Glycerine max* (L.) Merr.)

Sugar cane
(*Saccharum* sp.)

Sunflower
(*Helianthus annuus* L.)

Pastures

FIGURE 4.2 Distribution of main raw materials for biofuel productin in Brazilian territory.

FIGURE 4.3 Soybean cultivation in Mato Grosso state, Brazil.

tropical climates (Embrapa, 2006a). The seed of the soybean can generate many products and by-products for agricultural, chemical, and food uses. Seed production depends on external demand and the seeds are commonly used to produce a refined oil. Soybean oil represents 90% of total Brazilian production of vegetable oils and will serve as the basis for biodiesel production there for the short and medium terms. Because of the low yield of oil (around 18%), soybeans tend to be replaced gradually by other oil crops. This competition will decrease the cost of biodiesel production over time (Abiove, 2007).

It is worth noting that soybean may not be the best alternative for the production of biodiesel based on the cost of production of its oil when compared to other oilseeds. However, soybeans present two advantages for biodiesel: a well-structured production infrastructure and large-scale cultivation throughout Brazil. Biodiesel production from soybeans has great potential and may open new markets for traditional agriculture. However, this potential does not mitigate the negative environmental effects of soybean production on a large scale, especially if it is accompanied by an increase in cultivated area. Even if cultivated area does not increase, the traditional management system is not considered environmentally sustainable for biodiesel production. The intensive use of fertilizers leads to emission of N gases. Based on the final balance (expressed as CO_2 equivalents) throughout the biodiesel life cycle, soybean biodiesel would not reduce emissions more effectively than diesel oil.

Castor Bean

The castor bean (Figure 4.4) originated in Africa and is considered an oilseed with great potential for biodiesel production. It is easy to cultivate, economical to grow, and resistant to drought (Portal do Biodiesel, 2006). Through agricultural zoning in northeast Brazil, Embrapa Semi-Árido has already mapped 600,000 ha of land suitable for cultivation. Castor bean has a strong social component and is grown in consortium with other crops, particularly beans, by family farmers. One concern about castor bean cultivation is the sensitivity of the plant in competition with weeds. Weeds must be controlled to prevent decreasing yields of castor beans.

Castor bean oil is the main plant product; it is used in lubricants for turbines and manufacturing nylon, resins, fabrics, adhesives, and other products. The most important by-product of castor bean oil is castor bean cake—the residue from the extraction of oil from seeds. Biodiesel production from castor bean oil can represent great value in view of its high oil content (45% to 55%) and many uses of its by-products. Castor bean-planted area in Brazil is estimated at approximately 160,000 ha. Brazil is the third largest exporter of castor bean oil, accounting for about 12% of the world market and its main importing countries are the United States, Japan and countries of the European Union. The largest domestic producer is the state of Bahia, accounting for 92% of the harvest.

Regarding environmental benefits, each hectare cultivated with castor bean absorbs 10 t CO_2—four times the averages for other oilseeds. Even with these characteristics and more favorable incentives for the implementation of socioeconomic biodiesel plants in northeast Brazil, the productivity of this culture is very low in the region. Therefore, the promotion of castor bean cultivation must be accompanied by investments in agricultural research. Economic and technology barriers must be

FIGURE 4.4 Castor bean cultivation in Bahia state, Brazil.

overcome to add castor bean biodiesel to the Brazilian energy matrix (Freitas and Fredo, 2005).

Palm

The palm crop, known as "dendê" (Figure 4.5), is a perennial of African origin that arrived in Brazil in the 16th century and adapted to the southern coast of Bahia. Northern Brazil also exhibits favorable weather conditions for the cultivation of this oilseed. The state of Pará is the largest producer of palm oil, representing 80% of the cultivation area. When cultivated correctly, the palm harvest starts at the end of the third year; production ranges from 6 to 8 t of bunches ha^{-1}. The peak of the production occurs in the eighth year, reaching 25 t of bunches ha^{-1}. Production remains constant until the 17th year, declining slightly to the end of the plant's productive life—about 25 years (CEPLAC, 2007). Two types of oil are obtained from palm fruits: palm oil from the pulp or mesocarp, and the oil obtained from the kernel or endosperm (SUFRAMA, 2003). Palm oil is extracted by craft or industrial processes. Craft

FIGURE 4.5   Palm cultivation in Pará state, Brazil.

processes represent the majority of Bahia's oil processing units; industrial processes are used by medium or large companies.

Soil and climate conditions in southern Bahia and Pará are such that palm can be classified as a promising crop for biodiesel production. Although its oil concentration is low (22%), its high productivity per hectare allows the cultivation of this oilseed as a raw material for the biodiesel production. Among the advantages of palm oil are physical and chemical properties similar to diesel oil, good productivity, potential for large, scale production, and production throughout the year, thus requiring less space for storage and smaller processing facilities. Brazil has appropriate technology to increase the planted area with this perennial that produces up to 5 t of oil ha^{-1} yr^{-1}; it can be planted in areas already suited for other uses such as pastures (Furlan Júnior et al., 2004).

Sunflower

Sunflower (*Helianthus annuus* L.; Figure 4.6) is one of the major oilseeds in Brazil. It is suitable for biodiesel production due to excellent quality and high yield of oil extracted from seeds (45% to 55%). It is a rugged plant adapted easily to marginal soil and climatic conditions. In addition, it can be economically grown because it does not require specialized management.

The area planted with sunflower in Brazil covers approximately 76,000 ha and national production is about 115,000 t yr^{-1} (CONAB, 2006). Despite commercial scale production in the central west, southeast, and southern regions of Brazil, sunflower culture fits well throughout the country. It is also suitable for cultivation between crops. Used in rotation during the late season, sunflower could be grown

FIGURE 4.6   Sunflower cultivation. (*Source:* flickr.com/photos/ericoch/369871946/.)

on 20% of 13 million ha of soybean land and provide more than 2.5 billion liters of oil per year (Embrapa, 2006b). The high content of oil in the seeds—higher than other oilseeds—and the ease of extraction by pressing makes the sunflower suitable for small properties, also favoring the inclusion of family farmers in the production chain. The residue can be used for human and animal feed and other purposes.

Jatropha

Known as "pinhão manso" in Brazil (Figure 4.7), *Jatropha* is considered one of the most promising oilseeds to replace diesel oil; the kernels can generate 50% to 52% oil (Pinhão Manso, 2006). *Jatropha* is an oleaginous plant feasible for biodiesel because it produces at least 2 t ha^{-1} oil after the three to four years required to reach productive age, and this rate of production can continue for 40 years (Accarini, 2006). The plant probably originated in Brazil and was later introduced in the archipelago of Cape Verde and Guinea by Portuguese sailors and disseminated over the African continent. It is currently found in most tropical regions, growing on a large scale in tropical and temperate regions and, to a lesser extent, in cold regions (Brazil, 1985).

Jatropha has high productivity and is easily managed. Thus, it compares favorably with other oilseed crops. It is a perennial that grows throughout 90% of Brazilian territory. It can be grown in low fertility soils, different climates, and different soil types. These characteristics make it a strong candidate for inclusion in the vegetable oil program. Although its favorable characteristics for large-scale production are known, more studies are needed to enable large-scale production. Such studies should focus mainly on production issues because *Jatropha* was cultivated only as hedges on rural properties until recently.

FIGURE 4.7   Experimental *Jatropha* cultivation in Bahia state, Brazil.

BEEF TALLOW

Animal fats can also be used in biodiesel production. Biodiesel made from vegetable oils and from animal fats exhibit differences in physical properties. The main difference relates to cloud and melting points. A cloud point (CP) arises from a drop in temperature. The CP is defined as the temperature at which the formation of liquid crystals becomes visible and indicates solidification. Temperatures below CP cause crystal clustering that may restrict or prevent the free flow of fuel in pipes and filters, causing engine problems. The melting point (MP) is the temperature at which the crystal agglomerate is widespread enough to prevent the free flow of fluid (Knothe et al., 2006).

Biodiesel produced from animal fat has higher CP and MP than fuels derived from vegetable oils (Table 4.1). Thus, the formation of crystals is more likely to occur in tallow biodiesel, making its use in pure form more difficult. However, at same temperatures, vegetable and tallow biodiesels exhibit little difference in terms of pollutant emissions and engine performance (Van Gerpen, 1996).

The most feasible animal fat for biodiesel production in Brazil is bovine tallow. The country has the second largest cattle herd in the world, and tallow is presented as an alternative raw material for biodiesel with interesting availability and high potential for production. Tallow is a greasy residue composed of triglycerides that contain mainly palmitic (30%), stearic (20% to 25%), and oleic (45%) acids (Graboski and McCormick, 1998).

The potential for Brazil to produce biodiesel derived from beef tallow can be easily calculated based on available data from 2008. Considering that each animal

TABLE 4.1

Cloud and Melting Points of Tallow and Vegetable Biodiesel Fuels

Raw Material	Alkyl Group	Cloud Point ($^\circ$C)	Melting Point ($^\circ$C)
Canola	Methyl	−2	−9
Canola	Ethyl	−2	−15
Soybean	Methyl	0	−2
Soybean	Ethyl	1	−4
Sunflower	Methyl	2	−3
Sunflower	Ethyl	−1	−5
Beef tallow	Methyl	17	15
Beef tallow	Ethyl	15	12

Source: Knothe, G. et al. 2006. *Manual de Biodiesel.* Editora Blucher.

slaughtered provides 15 kg on average of serviceable tallow (RBB, 2006) and yields pure biodiesel equal to 80%, the slaughter of 22 million head in 2008 (IBGE, 2008) shows that the potential for production of biodiesel from beef tallow would be approximately 400 million liters.

The quality of tallow produced depends on the time of slaughter and the fat decomposition process. The actions of enzymes and bacteria cause changes in colors and free fatty acid content at death. Thus, enzymatic and bacterial control before and during slaughter is an essential factor for obtaining quality tallow. The efficiency of biodiesel production depends on quality, and therefore it is necessary to monitor the entire production cycle, from the animal slaughter through efficient sanitary control throughout production, transportation, and storage of the finished product.

MICROALGAE

In the search for alternative energy sources, microalgae can be considered potential raw materials for biodiesel. They are usually microscopic, prokaryotic, or eukaryotic, uni- or pluri-cellular, and photolitotrophic organisms. Among the photosynthetic organisms, microalgae are the most efficient in absorbing CO_2, and their growth is directly related to the reduction of GHGs since they require large quantities of CO_2 as carbon sources (Brown and Zeiler, 1993). Fatty acids and lipids are present in their cell membranes as well as in storage compounds, metabolites, and sources of energy (Banerjee et al., 2002). The production of microalgae biomass on a large scale is one possibility for providing oil supplies for biodiesel. Rich in lipids and fatty acids, the oil yield per hectare of some strains of microalgae is considerably higher than several conventional crops such as oil palm, *Jatropha*, soybean, and coconut (Table 4.2).

Oils found in microalgae cells show some physical and chemical properties similar to those of vegetable oils, and therefore, can be considered potential raw materials

TABLE 4.2
Possible Biodiesel Fuel Sources

Crop	Oil Yield (L ha⁻¹)	Land Area (M ha⁻¹)
Corn	172	1540
Soybean	446	594
Canola	1190	223
Jatropha	1892	140
Coconut	2689	99
Oil palm	5950	45
Microalgae*	136,900	2
Microalgae**	58,700	4.5

* 70% oil (by dry weight) in biomass.
** 30% oil (by dry weight) in biomass.
Source: Chisti, Y. 2007. Biotech. Adv. 25: 294–306.

for producing biodiesel (Chisti, 2007). Biodiesel from microalgae is a renewable energy source whose use does not contribute to the increase of GHGs in the atmosphere because production and use involve a closed cycle of CO_2. Moreover, the product is biodegradable, non-toxic, can be safely handled, and contains no sulfur, benzene, or other aromatic compounds (Wagner, 2007). The cultivation of microalgae provides a number of other economic advantages such as relatively low costs for harvesting and transportation, lower cost of water, use of infertile areas as supports for cultivation, high efficiency of CO_2 photosynthetic fixation by area, and the ability to grow in simple saline media (Danielo, 2005).

Current biotechnology studies using microalgae have gained attention (Sheehan et al., 1998; Dayananda et al., 2005; Spolaore et al., 2006). Microalgae cultivation as a source of renewable biomass can be applied to the production of biodiesel as a replacement for oil. Global warming is associated with the increases of carbon dioxide in the atmosphere (Chisti, 2007). The CO_2 resulting from industrial processes can also be used to cultivate microalgae, allowing CO_2 to serve as a raw material. The cultivation of microalgae requires large quantities of CO_2 nutrients (Benemann, 1997) with potential to function as a carbon sink.

Algal biomass can also be used for many purposes such as animal feeds, pigments, food supplements (Kay, 1991; Shimizu, 1996; Banerjee et al., 2002; Lorenz and Cysewski, 2003; Shimizu, 2003; Spolaore et al., 2006), and for bioremediation of contaminated areas (Kalin et al., 2005; Munoz and Guieysse, 2006). The main application of microalgae biomass has been in production of food supplements, and cultivation has been restricted to few species belonging to genera *Chlorella*, *Dunaliella*, *Scenedesmus*, and *Spirulina* (Becker, 2004).

Consumption of microalgae for food is a very old tradition. Native peoples from Asia consumed some species of genus *Nostoc*, and other people such as the Aztecs in Mexico and the Kanembous in Africa consumed algae of genus *Spirulina*. The winds pushed and clustered algae biomasses; the biomasses were dried, mashed, and cut into small slabs (Durand-Chastel, 1980; Dillon et al., 1995). The commercial production of microalgae triggered strong interest starting at the 1960s, with the development of technologies to produce biomass for cultivation of microalgae in open ponds and photobioreactors (Wagner, 2007).

The use and advantages of various genera of microalgae for the production of biodiesel have been widely discussed in the literature (Kawaguchi, 1980; Hillen et al., 1982; Sheehan et al., 1988; Borowitzka, 1998; Sawayama et al., 1999; Chisti, 2007; Chisti, 2008; Reijnders, 2008). Among the genera reported, *Botryococcus* has been described as one of the most efficient producers of oil (Ranga Rao et al., 2007; Table 4.3). *Botryococcus braunii* is a green colonial microalgae found in lakes and reservoirs of fresh and brackish water. This species has attracted great scientific and commercial interest because of its ability to accumulate high amounts of lipids that can be converted into biodiesel, jet fuel, gasoline, and other important chemicals (Metzger and Largeau, 2005).

Species of *Botryococcus* produce various types of hydrocarbons, of which botryococens are the most important because they are produced in large quantities (Dayananda et al., 2005). Depending on the type of hydrocarbon produced, *B. braunii* (Figure 4.8) is classified among strains A, B, or L. Algae belonging to strain A produce primarily oil and n-alkadiene and triene hydrocarbons designated C_{23}–C_{33} (Metzger et al., 1985); strain B produces triterpenoid hydrocarbons (C_{30}–C_{37}) botryococens (Metzger et al., 1985) and methyl squalene (C_{31}–C_{34}) (Achitouv et al., 2004), while strain L produces a single type of triterpenoid hydrocarbon, named licopedien

TABLE 4.3
Oil Contents of Some Species of Microalgae

Species	Oil Content (% Dry Weight)
Botryococcus braunii	25 to 75
Chlorella sp.	28 to 32
Crypthecodinium cohnii	20
Cylindrotheca sp.	16 to 37
Dunaliella primolecta	23
Isochrysis sp.	25 to 33
Monallanthus salina	>20
Nannochloris sp.	20 to 35
Nannochloropsis sp.	31 to 68
Neochloris oleoabundans	35 to 54
Nitzchia sp.	45 to 47
Phaeodactylum tricornutum	20 to 30
Schizochytrium sp.	50 to 77
Tetraselmis sueica	15 to 23

Source: Chisti, Y. 2007. *Biotech. Adv.* 25: 294–306.

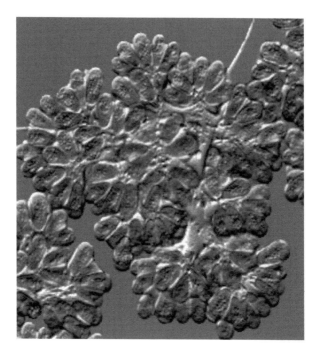

FIGURE 4.8 Morphology of *Botryococcus braunii*. (*Source:* runcesab.unblog.fr/micro-algues/botryococcus-braunii/.)

(Metzger et al., 1990). *B. braunii* species also synthesize fatty acids, triacil glycerols, and sterols (Dayananda et al., 2005).

The U.S. National Renewable Energy Laboratory (NREL) and Department of Energy Office of Fuels Development (DOE) were pioneers in 1970 that started researching strains of microalgae for use in biofuels, based on possible oil shortages (Wagner, 2007). From 1978 to 1996, DOE established a program for the sustainable production of oil from algae. Since then, the main focus of the Aquatic Species Program (ASP) has been the production of biodiesel from pond algae exhibiting high contents of lipids, utilizing waste CO_2 from coal burning plants. Located in Golden, Colorado, the NREL is a consortium of research laboratories aiming to produce oil from microalgae. NREL has selected about 300 species of marine and freshwater microalgae, mainly diatoms (genera *Amphora*, *Cymbella*, *Nitzsche*, etc.) and green algae (especially genus *Chlorella*) (Danielo, 2005).

According to NREL scientists, the oil yield from algae biomass is at least 30 times higher than yields from oil crops such as palm, sunflower, and peanut, commonly used to produce oil. The ability of microalgae to grow in liquid media allows them greater access to water, CO_2, and nutrients. Another important issue is the surface area exposed to the sun rather than the volume occupied. Thus microalgae productivity is measured in terms of biomass (kilograms of algae or oil) per day, per unit area exposed to the sun. This allows comparisons with data from terrestrial plants (Danielo, 2005).

LAND USE FOR OILSEED CULTIVATION

Competition with Food Crops

When compared to other countries, Brazil has advantages for renewable energy production and consequently a high potential for supplying the global energy matrix. One reason that led Brazil to this advantageous position is the prospect of incorporating raw material production areas. Of course, production systems should be organized so as not to compete with food agriculture. Data from IBGE (2006) show a decrease between 1995 and 2006 in the average cultivated areas for basic crops of the Brazilian diet such as beans and rice. The data also show increases in soybean and sugar cane production—the main crops used to generate renewable energy in Brazil.

Despite these data, an increase in food crop productivity has been observed. Thus, it appears that sugar cane and soybean production have not competed with basic crops such as rice and beans. This increase in the production of food crops results from improvements in productivity and regional specialization processes. Over the past 10 years, food crops have released cultivation areas without pressure from the expansion of energy agriculture.

Brazil can cultivate multiple crops within a calendar year. The agricultural system encompasses cultivation in the late season, summer double cropping, and winter cultivation. This system is commonly used for the production of grains. The model is called "productive windows" and consists of periods posing reasonable risks for main crops, with acceptable risks to other crops, allowing possibilities for energy agriculture (Hernandez, 2008).

Cultivated Area Expansion and High Productivity of Oilseed Production

The wide range of oilseed crops grown for biodiesel production in Brazil led to exploration of other aspects of agriculture not related to competition with food crops. One issue is use of areas that are marginal or unsuitable for food crop cultivation to produce oilseed species. An estimate of available areas for expansion of oilseed crops in Brazil (CONAB, 2006) showed the capacity to produce biodiesel from oils of soybeans, castor beans, palms, and sunflowers (Table 4.4). Palm, a high-yield potential crop, has smaller expansion areas than soybean and castor bean. Due to its high annual productivity, oil production may be at least twice the production of other oilseeds considered in the estimate. Table 4.5 presents data on oil yield, productivity, and estimated expansion needed to achieve 5 million tonnes of oil per year in Brazil.

The examined cultures have low annual productivity per hectare, even considering the high oil content of castor bean and sunflower. Among these cultures, palm deserves attention. Despite its low percentage of oil, it is a perennial with high yield potential, requiring less expansion area than other oilseeds. The high productivity and the low cost of production make palm oil competitive in the world market for biodiesel production compared to other oil seeds.

The Brazilian Amazon region has about 70 million ha with potential for cultivation of palm. A small fraction has been effectively used for palm cultivation, and

TABLE 4.4
Estimated Planted and Expansion Areas and Oil Production Capacity of Main Oilseeds Used for Biodiesel Production

Crops	Planted Area (1,000 ha)	Expansion Area (1,000 ha)	Oil Production Capacity (t ha^{-1}yr^{-1})
Soybean	22,213.1	1,090.4	436,160
Castor bean	153.0	881.4	661,050
Palm	88.7*	471.7	1,415,100
Sunflower	38.5	121.9	97,520
Total		2,565.4	2,609,830

* For 2005.

Source: Companhia Nacional de Abastecimento. 2006. Girassol: comparativo de área, produtividade e produção. http://www.conab.gov.br/conabweb/download/cas/semanais/semana15a19052006/conj_girassol_11_a_15_09_06.pdf (accessed April 2008).

TABLE 4.5
Estimated Expansion Areas for Biodiesel Oilseed Cultivation in Brazil

Crops	Oil Yield (%)	Productivity (t ha^{-1} yr^{-1})	Oil Extraction (t ha^{-1}yr^{-1})	Extraction (1 t oil) Area (ha)	Extraction (5,000 t oil) area (1,000 ha)
Soybean	18	2.2	0.40	2.50	12,500.0
Castor bean	50	1.5	0.75	1.33	6,650.0
Palm	20	15	3.00	0.33	1,650.0
Sunflower	50	1.6	0.80	1.25	6,250.0

about 85% of the cultivated area is located in the state of Pará (Embrapa, 2002). The main soil and climate conditions of the Amazon region such as high rainfall, little or no drought, and deep soils make palm crops well adapted (SUFRAMA, 2003). Moreover, palm can be grown in concert with leguminous crops and cultivated in areas where soils are partially degraded, allowing land recovery and thus contributing to the sustainable development of the Amazon.

USE OF IDLE LANDS FOR FOOD CROPS

Within the context of land use, certain areas that are poorly adapted for production of food crops may be directed to the production of energy crop species that are suitable for regional conditions. The semi-arid northeast region of Brazil merits attention. It is an ecologically fragile area that does not support production systems based on traditional models of agriculture. The excessive deforestation followed by techniques that did not preserve or recover soil structures was the first step to trigger the desertification process and exhibit dangerous impacts for the medium and long terms.

One prominent issue related to the semi-arid northeast region is the use of soil and water conservation practices among the local farmers to guarantee the agricultural productivity of land. Among the alternatives considered for improving production systems while conserving soil environments are the replacement of burning with other practices, plantation in level curves, preservation of soil moisture near plants, and rotations with crops that supply and fix nitrogen in the soil.

Castor bean, a nitrogen-fixing oilseed, has been cultivated in the semi-arid northeast and shows certain characteristics such as strong resistance to drought and high requirements for heat and light (Amorim Neto et al., 1999). The semi-arid rainfall is associated with high levels of agricultural productivity of castor bean (Beltrão, 2004). Castor bean cultivation occupies a prominent place in the program of biodiesel production because it has aroused great interest for family agriculture. Currently Bahia state is the largest producer of castor bean, responsible for 92% of Brazilian production. In addition to economic benefit, castor bean is a major source of biomass and energy and can sequester around 10 t ha^{-1} C, thereby helping reverse global warming (Beltrão, 2004).

A very promising alternative for low fertility soils in semi-arid regions is the cultivation of *Jatropha*. This plant is under study with the aim of helping meet the ever-increasing demands for biodiesel production (Lima, 2007). *Jatropha* exhibits great potential for oil production based on a unique feature that differentiates it from other oil crops: Its production cycle exceeds 40 years.

The oil yield of *Jatropha* is less than that of castor bean; however, because planting is required only once in 40 years, production costs are greatly reduced (Albuquerque, 2008). Moreover, *Jatropha* has a hardy root system that makes it resistant to drought. It can grow in environments that get as little as 200 mm of precipitation annually and can withstand up to three consecutive years of drought (Saturnino et al., 2005). Based on these characteristics and the ability to develop in soils of low fertility, *Jatropha* shows potential for integration into the biodiesel production program in the semi-arid region of northeast Brazil (Arruda et al., 2004).

Besides encouraging the cultivation of castor beans, the Brazilian government and private sector have developed several initiatives to disseminate information about the production system of *Jatropha* in semi-arid conditions. Research on oil crop agriculture in the region is still very incipient and it is still necessary to search for technical information to promote sustainable production systems.

SOIL CARBON SEQUESTRATION IN PRODUCTION OF BIODIESEL

From an environmental view, the first justification to use biodiesel as a substitute for diesel oil is the possibility of having a CO_2-neutral system. The hypothesis is that all the CO_2 emitted from burning is absorbed by photosynthesis, but this view fails to consider energy inputs required to plant, grow, harvest, transport, process, and distribute fuels and the release of CO_2 from burning of biodiesel. Consequently, the degree to which any biodiesel may decrease CO_2 emissions in comparison with fossil oil depends on production and refining methods (Powlson et al., 2005; Turner et al., 2007).

Development of biofuels may increase CO_2 emissions significantly if forested lands are cleared for energy crops (Giampietro et al., 1997; Junginger et al., 2006). As Brazilian forests were converted first to pastures and then to sugarcane plantations, soil organic carbon was depleted by more than 40% (Silveira et al., 2000). Overall, forest conversion accounted for 75% of Brazil's GHG from 1990 to 1994 (Macedo et al., 2004).

Similarly, oil palm plantations have spread extensively in Malaysia, Indonesia, and Thailand, replacing tropical forest habitats and releasing stored C as forests are cleared and soils exposed. If oil palm plantations are sited on deep peat soils, CO_2 emissions from the soil can be substantial (Worldwatch Institute, 2006). Thus, the potential for biofuel production to mitigate climate warming depends on the types of lands used for biofuel crop cultivation (Smeets et al., 2005; Marshall, 2007).

Intensive land use exerts negative effects on the environment and agricultural productivity when conservation practices are not adopted (Cerri et al., 2004; Foley et al., 2005). The reduction of soil organic matter is accompanied by increased release of GHGs to the atmosphere and increases global warming (Knorr et al., 2005). Intensive land use reduces the quality of organic matter remaining in the soil. These changes occur, for example, through the breakdown and destruction of soil with losses to erosion, reducing the availability of nutrients to plants and decreasing water storage capacity. These are some factors that reflect negatively on agricultural productivity and consequently on food production and the sustainability of the soil–plant–atmosphere system (Lal, 2003; Six et al., 2004; Knorr et al., 2005).

Soil organic carbon (SOC) sequestration is affected by crop management decisions that impact the quantity and quality of crop residue added to the soil and the rate of decomposition (Paustian et al., 2000; Jarecki and Lal, 2003). Crops have different requirements for energy inputs during planting, tillage, fertilizer and pesticide application, and harvest (West and Marland, 2002).

CARBON SEQUESTRATION POTENTIAL OF PASTURE LAND

Estimating pasture C sequestration potential is more difficult than making such estimates for agricultural areas. Pastures include a wide diversity of plants and soils. Moreover, management practices may induce changes in the whole community of plants and may in the long term exert secondary effects on soil C stock (Schuman et al., 2002).

Conversion of pasture to crops results in carbon storage on the order to 0.5 Mg C ha^{-1} yr^{-1} on average for 50 years (IPCC, 2000). This soil C increase could be considered low and shows that even after 50 years soil C does not recover to the level prior to the introduction of the first pasture. Due to the low C accumulation rate, it is considered that pastures used more than 20 years do not achieve C sequestration (Franck, 2000).

The increase of soil C after conversion of crops to pasture (in the absence of frequent soil tillage) can be explained mainly by higher C supply to soil (through roots and shoots), the increased time that carbon stays in the system due to lack of tillage, and continuous protection of soil organic matter (SOM) that reduces losses (Soussana et al., 2004). Soil tillage reduces the physical protection of organic matter, resulting in humic fraction reduction (Post and Kwon, 2000). Furthermore, soil tillage increases aggregate degradation, accelerating organic matter decomposition (Paustian et al., 2000), consequently reducing C content.

Organic matter in soils under pasture is rich in aromatic compounds and thus exhibits greater resistance to degradation (Gregorich et al., 2001). After pasture introduction, roots and microflora tend to stabilize soil aggregates (Jastrow, 1996). Three reasons (Balesdent et al., 2000) may explain higher pasture C sequestration potential compared to crops:

- A large proportion of SOM comes from pastures, root exudation, and decomposition that provide protection in the form of particulate organic matter.
- Much of the particulate SOM is chemically stable.
- Soil aggregates tend to protect SOM from decomposition.

Consequences of Conversion of Forest to Pasture

In Brazil, deforestation is a function mainly of opening new areas for the introduction of pastures. Native forest removal and pasture introduction have impacted soil C content as has the conversion of forests into pastures resulting from reforestation initiatives or natural reforestation on abandoned degraded pastures

Conversion of forest to pasture can lead to C accumulation or release, depending on soil conditions (Post and Kwon, 2002). The introduction of pasture usually decreases SOM levels during the first years of implementation, followed by increasing levels that approach previous levels in primary forest (Chonè et al., 1991; Resck et al., 2000). This increase can be significant to the point of achieving small gains in organic matter compared to natural ecosystems (Corazza, 1999). In natural ecosystems, the soil organic carbon has a single origin—plant residues of native vegetation. In agrosystems, most of soil carbon comes from at least two sources: decomposition of native vegetation and introduced crops (Bernoux et al., 1999).

Reforestation areas also exhibit great C sequestration potential, but differently from pasture. The pasture-to-forest conversion presents significant change in the nature and location of crop residues to be incorporated into soil. In pastures, organic carbon input to soil occurs via the dense mass decomposition of fine and deep roots. In forests, carbon input occurs mainly through deposition of shoots from trees that remains on the soil surface, hindering carbon incorporation to SOM (Lima et al., 2006).

Grasslands and forests are complex ecosystems, and it is very difficult to compare the C sequestration potentials of these ecosystems because data discrepancies appear in the literature. According to Davis and Domanski (2002), soil organic matter reduction occurs after pasture-to-forest conversion in the short term, but recovers in the long term. Mendham et al. (2002) found similar carbon stocks in well-managed pastures and forests, while Lima et al. (2006) found soil carbon increasing with conversion of degraded pastures to forests.

However, even with inconsistent results for degraded pastures, introduction of forests provides unquestionable benefits. Abandonment of degraded areas, as occurs in some regions of Brazil, is a land waste that promotes further deforestation and soil organic carbon losses. Such areas could be reused and appropriately managed with pastures or crops, thereby promoting favorable conditions for C sequestration in soil. Keep in mind that C sequestration capacity in any soil type depends on its physical characteristics (especially micro-aggregation), chemical (content of silt and clay) and organic (recalcitrant humic substances formation) (Six et al., 2001).

FEASIBILITY OF CARBON SEQUESTRATION IN SYSTEMS OF OIL CROP CULTIVATION

When considering the life cycle of an oilseed crop, it appears that the agricultural stage is the most questioned since it has direct relationship with C sequestration in the soil. The collection of such data is important to demonstrate the environmental feasibility of replacing diesel oil with biodiesel. The use of sustainable productive systems without burning and with effective soil and water conservation practices appears a viable alternative for the cultivation of oilseed species. In agricultural practice, several actions should be implemented to reduce GHG emissions and promote C sequestration in soil. Among them, we can highlight technologies applied to systems of oil crop production, with emphasis on adopting conservationist practices of soil management such as no-tillage and minimum tillage.

When conventional tillage is used for soil preparation or after harvest, part of the carbon is removed from soil or retained by plants and part is lost to the atmosphere. The intensive soil movements promote losses of soil carbon as CO_2 and also contribute to increasing emissions of CH_4 and N_2O to the atmosphere. In farming systems involving less intensive cultivation practices such as no-tillage, GHG emissions are much lower due to retention of such gases in the soil. In addition to the increase of soil organic matter contents, the adoption of such conservation practices reduces the traffic of machines on the plots, attenuating the emissions of CO_2 derived from burning machinery fuel to the atmosphere.

Different land uses can contribute to carbon mitigation. In addition to conservation practices, production of perennial crops such as *Jatropha* and palm can be adopted. Because such systems do not promote constant turning of the soil, the oil crops store large amounts of carbon in their biomasses, part of which can be returned to the soil through crop residues.

Several studies have evaluated the energy balances (Marland and Turhollow, 1991; Shapouri et al., 2002; Farrell et al., 2006) and fluxes of GHGs (Sheehan et al., 1998, 2004; McLaughlin et al., 2002; Heller et al., 2004; Updegraff et al., 2004; Kim and Dale, 2005) of specific crops for bioenergy, but information about comparisons of various crops is limited (Kim and Dale, 2004). It is necessary to integrate factors that contribute to the impacts of land use changes in the fluxes of GHGs.

Studies examining GHG emissions and carbon stocks throughout the agricultural chains of production of oil crops (soil preparation, fertilization, sowing, application of pesticides, management, and harvesting) are required to determine whether C sequestration actually occurs in productive systems. If the results are positive, they will confirm that biodiesel is an environmentally correct substitute for diesel oil.

FINAL CONSIDERATIONS

Currently, in the context of climate change, biofuel use as a substitute for fossil fuels can be considered the main mitigation measure. In terms of structure of the biodiesel production chain, soil management takes an important role, mainly through sustainable practices and conservation management. In Brazil, the largest sources of GHG emissions are deforestation and land use changes followed by fossil fuel burning. The country is described as a future leader in biodiesel production because of its excellent soil and climate conditions and vast land area. Diverse oilseed crops can be cultivated in Brazilian territory without promoting competition with food crops and without increasing deforestation. Moreover, variable raw materials such as animal tallow may serve as viable alternatives based on large production volumes, avoiding its disposal in the environment and adding economic value to such products.

Studies of land use are required to assess the environmental impact of the biodiesel agricultural chain. Besides GHG emissions, comparative analyses of biodiesel and diesel oil are also necessary to analyze soil C sequestration in the production stages of different oilseeds. The use of animal tallow as a raw material requires appropriate pasture management that will provide high C sequestration values and help reduce climate change.

Within an environmental context, further studies are needed along with investments that contribute to the use of raw materials with high yield potentials that require minimum natural resources such as soil and water. Oil seed crops such as *Jatropha* that can be adapted to low soil fertility and water-deficient conditions should be considered. The cultivation of microalgal biomass for production of biodiesel has attracted considerable scientific interest because of its potential for environmentally sustainable production that can replace diesel derived from fossil fuels (about 40 billion liters annually).

ACKNOWLEDGMENTS

We thank Petróleo Brasileiro S/A (Petrobras), Fundação de Apoio à Pesquisa do Estado de São Paulo (FAPESP), and Coordenação de Aperfeiçoamento de Pessoal de Nível Superior (CAPES) for financial support.

REFERENCES

Accarini, J.H. 2006. Biodiesel no Brasil: Estágio atual e perspectivas. *Bahia Anal. Dados,* 16: 51–63.

Achitouv, E. et al. 2004. C_{31}–C_{34} methylated squalenes from a Bolivian strain of *Botryococcus braunii. Phytochemistry* 65: 3159–3165.

Adler, P.R., S.J. Del Grosso, and W.J. Parton. 2007. Life cycle assessment of net greenhouse gas flux for bioenergy cropping systems. *Ecol. Appl.* 17: 675–691.

Albuquerque, B.C.D. 2008. O uso do pinhão-manso no semi-árido potiguar: Uma alternativa sustentável III. Congresso de Pesquisa e Inovação da Rede Norte Nordeste de Educação Tecnológica. Fortaleza, CE.

Amorim Neto, M.S. et al. 1999. Zoneamento e época de plantio para mamoneira no Estado da Bahia. Circular Técnica 103. Campina Grande: Empresa Brasileira de Pesquisa Agropecuária.

Arruda, F.P. et al. 2004. Cultivo de pinhão manso (*Jatropha curca* L.) como alternativa para o semi-árido nordestino. Revisão. *Rev. Bras. Ol. Fibros.* 8: 789–799.

Associação Brasileira das Indústrias de Óleos Vegetais Abiove. 2007. Produção responsável no agronegócio soja. São Paulo, Brazil.

Balesdent, J., C. Chenu, and M. Balabane. 2000. Relationship of soil organic matter dynamics to physical protection and tillage. *Soil Till. Res.* 53: 215–230.

Banerjee, A. et al. 2002. *Botryococcus braunii*: A renewable source of hydrocarbons and other chemicals. *Crit. Rev. Biotech.* 22: 245–279.

Batjes, N.H. 1999. Management options for reducing CO_2 concentrations in the atmosphere by increasing carbon sequestration in the soil. Technical Paper 30. ISRIC.

Becker, E.W. 2004. Microalgae in human and animal nutrition. In Richmond, A., Ed., *Handbook of Microalgal Culture: Biotechnology and Applied Phycology*. London: Blackwell, pp. 312–351.

Beltrão, N.E.M. 2004. A cadeia da mamona no Brasil, com ênfase para o segmento P&D. Documentos 129. Campina Grande: Empresa Brasileira de Pesquisa Agropecuária.

Bernoux, M. et al. 1999. Carbono e nitrogênio em solo de uma cronossequência de floresta tropical: Pastagem em Paragominas, Pará, Brasil. *Sci. Agric.* 56: 777–783.

Borowitzka, M.A. 1998. Fats, oils and hydrocarbons. In Borowitzka, M.A. et al., Eds., *Microalgal Biotechnology*. Cambridge University Press, pp. 257–287.

Brazil. Ministério da Indústria e do Comércio. Secretária de Tecnologia Industrial. 1985. Produção de combustíveis líquidos a partir de óleos vegetais. Documentos 16. Brasília, STI/CIT.

Brown, L.M. and K.G. Zeiler. 1993. Aquatic biomass and carbon dioxide trapping. *Energ. Conv. Manag.* 34: 1005–1013.

Carnielli, F. 2003. O combustível do futuro. Boletim informativo, 1413. www.ufmg.br/boletim/bul1413/quarta.shtml (accessed January 2008).

Cerri, C.C. and F.G. Andreux. 1990. Changes in organic carbon content in oxisols cultivated with sugar cane and pastures based on ^{13}C natural abundance measurement. In International Congress Of Soil Science, 14, Kyoto.

Cerri, C.E.P. et al. 2004. Modeling changes in soil organic matter in Amazon forest to pasture conversion with Century model. *Global Change Biol.* 10: 815–832.

Chisti, Y. 2007. Biodiesel from microalgae: Research review. *Biotech. Adv.* 25: 294–306.

Chisti, Y. 2008. Biodiesel from microalgae beats bioethanol. *Trends Biotech.* 26: 126–131.

Chonè, T. et al. 1991. Changes in organic matter in an Oxisol from the central Amazonian forest during eight years as pasture, determined by ^{13}C isotopic composition. In Berthelin, J., Ed., *Diversity of Environmental Biogeochemistry*. Amsterdam: Elsevier, pp. 397–405.

Comissão Executiva do Plano da Lavoura Cacaueira. 2007. Dendê: Potencial para produção de energia renovável. http://www.ceplac.gov.br/radar/ Artigos/artigo9.htm (accessed March 2008).

Companhia Nacional de Abastecimento. 2006. Girassol: Comparativo de área, produtividade e produção. http://www.conab.gov.br/conabweb/download/cas/semanais/semana15a19052006/conj_girassol_11_a_15_09_06.pdf (accessed April 2008).

Corazza, E.J. et al. 1999. Comportamento de diferentes sistemas de manejo como fonte ou depósito de carbono em relação a vegetação do cerrado. *Rev. Bras. Ci. Solo* 23: 425–432.

Danielo, O. 2005. An algae-based fuel. *Biofutur* 255: 1–4.

Davis, M.R. and L.M. Condrom. 2002. Impact of grassland afforestation on soil carbon in New Zealand: A review of paired site studies. *Austr. J. Soil Res.* 40: 675–690.

Dayananda, C. et al. 2005. Effect of media and culture conditions on growth and hydrocarbon production by *Botryococcus braunii*. *Proc. Biochem.* 40: 3125–3131.

Del Grosso, S.J. et al. 2001. Simulated interaction of carbon dynamics and nitrogen trace gas fluxes using the DAYCENT model. In Schaffer, M. et al., Eds., *Modeling Carbon and Nitrogen Dynamics for Soil Management.* Boca Raton, FL: CRC Press, pp. 303–332.

Demattê, J.L.I. 1998. *Manejo de solos ácidos dos trópicos úmidos: Região Amazônica.* Ed. Fundação Cargill.

Dillon, J., A.P. Phuc, and J.P. Dubacq. 1995. Nutritional value of the alga *Spirulina*. *Plants Hum. Nutr.* 77: 32–46.

Durand-Chastel, H. 1980. Production and use of *Spirulina* in Mexico. In Shelef, G. and C.J. Soeder, Eds., *Algae Biomass.* Amsterdam: Elsevier, pp. 51–64.

Eden, M.J. et al. 1991. Effect of forest clearance and burning on soil properties in northern Roraima, Brazil. *Forest Ecol. Mgt.* 38: 283–290.

Empresa Brasileira de Pesquisa Agropecuária. 2002. Dendê Alternativa de desenvolvimento sustentável para agricultura familiar Brasileira. Boletim Técnico. Manaus.

Empresa Brasileira de Pesquisa Agropecuária. 2006a. A cultura da soja. http://www.cpso. embrapa.br/ (accessed March 2006).

Empresa Brasileira de Pesquisa Agropecuária. 2006b. Girassol no Brasil. Londrina, 2006.

Farrell, A.E. et al. 2006. Ethanol can contribute to energy and environmental goals. *Science* 311: 506–508.

Fearnside, M.P. 1996. Amazon deforestation and global warming: Carbon stocks in vegetation replacing Brazil's Amazon forest. *Forest Ecol. Mgt.* 80: 21–34.

Foley, J.A. et al. 2005. Global consequences of land use. *Science* 309: 570–574.

Frank, A.B. 2002. Carbon dioxide fluxes over a grazed prairie and seeded pasture in the Northern Great Plains. *Environ. Poll.* 116: 397–403.

Freitas, S.M. and C.E. Fredo. 2005. Biodiesel a base de óleo de mamona: Algumas considerações. *Inform. Econ.* 35: 37–42.

Furlan Júnior, J. et al. 2004. A utilização de óleo de palma como componente do Biodiesel na Amazônia. *Comun. Técn.* 103.

Giampietro, M., S. Ulgiati, and D. Pimentel. 1997. Feasibility of large-scale biofuel production: Does an enlargement of scale change the picture? *BioSci.* 47: 587–600.

Goldemberg, J. 2007. Ethanol for a sustainable energy future. *Science* 315: 808–810.

Graboski, M.S. and R.L. McCormick. 1998. Combustion of fat and vegetable oil derived fuels in diesel engines. *Prog. Energy Comb. Sci.* 24: 125–164.

Gregorich, E.G., C.F. Drury, and J.A. Baldock. 2001. Changes in soil carbon under long-term maize in monoculture and legume-based rotation. *Can. J. Soil Sci.* 81: 21–31.

Groom, M.T., E. M. Gray, and P.A. Townsend. 2008. Biofuels and biodiversity: Principles for creating better policies for biofuel production. *Conserv. Biol.* 22: 602–609.

Heller, M.C. et al. 2004. Life cycle energy and environmental benefits of generating electricity from willow biomass. *Renew. Energy* 29: 1023–1042.

Hernandez, D.I.M. 2008. Efeitos da produção de etanol e biodiesel na produção agropecuária do Brasil. Universidade de Brasília: Brasília.

Hillen, L.W. et al. 1982. Hydrocracking of the oils of *Botryococcus braunii* to transport fuels. *Biotech. Bioeng.* 24: 193–205.

Holanda, A. 2004. Biodiesel e Inclusão Social: Relatório apresentado ao Conselho de Altos Estudos e Avaliação Tecnológica. Centro de Documentos e Informação Coordenação de Publicação. Caderno de Altos Estudos 1. Brasília.

Instituto Brasileiro de Geografia e Estatística. 2006. Censo Agropecuário: Resultados Preliminares.

Instituto Brasileiro de Geografia e Estatística. 2008. Levantamento Sistemático da Produção Agrícola.

Instituto Nacional de Pesquisas Espaciais. 2000. Monitoramento da Floresta Amazônica por Satélite 1998–1999. http://www.inpe.br (accessed February 2009).

Intergovernmental Panel on Climate Change. 2000. *Land Use, Land Use Change and Forests.* Special Report, Cambridge.

Jarecki, M.K. and R. Lal. 2003. Crop management for soil carbon sequestration. *Crit. Rev. Plant Sci.* 22: 471–502.

Jastrow, J.D. 1996. Soil aggregate formation and the accrual of particulate and mineral-associated organic matter. *Soil Biol. Biochem.* 28: 665–676.

Judd, B. 2002. Biodiesel from tallow. Reports database prepared for Energy Efficiency and Conservation Authority.

Junginger, M. et al. 2006. The growing role of biofuels: Opportunities, challenges, and pitfalls. *Int. Sugar J.* 108: 615–629.

Kalin, M., W.N. Wheeler, and G. Meinrath. 2005. The removal of uranium from mining waste water using algal/microbial biomass. *J. Environ. Rad.* 78: 151–77.

Kawaguchi, K. 1980. Microalgae production systems in Asia. In *Algal Biomass: Production and Use.* Amsterdam: Elsevier, pp. 25–33.

Kay, R.A. 1991. Microalgae as food and supplement. *Crit. Rev. Food Sci. Nutr.* 30: 555–573.

Kim, S. and B.E. Dale. 2004. Cumulative energy and global warming impact from the production of biomass for biobased products. *J. Ind. Ecol.* 7: 147–162.

Kim, S. and B.E. Dale. 2005. Environmental aspects of ethanol derived from no-tilled corn grain: Nonrenewable energy consumption and greenhouse gas emissions. *Biom. Bioenergy* 28: 475–489.

Knorr, W. et al. 2005. Long-term sensitivity of soil carbon turnover to warming. *Nature* 433: 298–301.

Knothe, G. et al. 2006. *Manual de Biodiesel.* Editora Blucher.

Lal, R. 2003. Global potential of soil carbon sequestration to mitigate the greenhouse effect. *Crit. Rev. Plant Sci.* 22: 151–184.

Lima, A.M.N. et al. 2006. Soil carbon dynamics following afforestation of degraded pastures with eucalyptus in southeastern Brazil. *Forest Ecol. Mgt.* 235: 219–231.

Lima, P.C.R. 2007. O biodiesel no Brasil e no mundo e o potencial do Estado da Paraíba. Estudo Técnico da Consultoria Legislativa da Câmara dos Deputados.

Lorenz, R.T. and G.R. Cysewski. 2003. Commercial potential for *Haematococcus* microalga as a natural source of astaxanthin. *Trends Biotechnol.* 18: 160–167.

Macedo, I.C., M.R. Leal, and J.E.R. Silva. 2004. Greenhouse gas emissions and energy balances in bioethanol production and use in Brazil. Secretaria do Meio Ambiente, São Paulo, Brazil. ww.unica.com.br/i pages/files/gee3.pdf (accessed April 2007).

Macedo, M.C.M. 2000. A integração lavoura e pecuária como alternativa de recuperação de pastagens degradadas. p. 90-104. In *Workshop Nitrogênio na Sustentabilidade de Sistemas Intensivos de Produção Agropecuária.* Embrapa Agropecuária Oeste; Seropédica: Embrapa Agrobiologia.

Marland, G. and A.F. Turhollow. 1991. CO_2 emissions from the production and combustion of fuel ethanol from corn. *Energy* 16: 1307–1316.

Marshall, L. 2007. Thirst for corn: What 2007 plantings could mean for the environment. Policy note. World Resources Institute, Washington, D.C. www.wri.org/ policynotes (accessed June 2007).

McLaughlin, S.B. et al. 2002. High-value renewable energy from prairie grasses. *Environ. Sci. Technol.* 36: 2122–2129.

Mendham, D.S., A.M. Connell, and T.S. Grove. 2002. Organic matter characteristics under native forest, long-term pasture, and recent conversion to eucalyptus plantations in Western Australia: Microbial biomass, soil respiration and permangate oxidation. *Austr. J. Soil Sci.* 40: 859–872.

Metzger, P. et al. 1990. Structure and chemistry of a new chemical race of *Botryococcus braunii* that produces lycopadiene, a tetraterpenoid hydrocarbon. *J. Phycol.* 26: 258–266.

Metzger, P., and C. Largeau. 2005. *Botryococcus braunii*: A rich source for hydrocarbons and related ether lipids. *Appl. Microbiol. Biotech.* 66: 486–496.

Metzger, P. et al. 1985. Alkadiene and botryococcene producing races of wild strains of *Botryococcus braunii*. *Phytochem.* 24: 2305–2312.

Munoz, R. and B. Guieysse. 2006. Algal-bacterial processes for the treatment of hazardous contaminants: A review. *Water Res.* 40: 2799–2815.

Nepstad, D., C. Uhl, and E.A. Serrão. 1990. Surmounting barriers to forest regeneration in abandoned, highly degraded pastures: A case study from Paragominas, Pará, Brasil. In Anderson, A.B., Ed. *Alternatives to Deforestation: Steps toward Sustainable Use of the Amazon Rain Forest*. Columbia University Press, pp. 215–229.

Pacala, S. and R. Socolow. 2004. Stabilization wedges: Solving the climate problem for the next 50 years with current technologies. *Science* 305: 968–972.

Paustian, K., J. Six, E.T. Elliott, and H.W. Hunt. 2000. Management options for reducing CO_2 emissions from agricultural soils. *Biogeochemistry* 48: 147–163.

Pinhão Manso. 2006. Pinhão Manso: Uma planta do futuro. http://www.pinhaomanso.com.br (accessed February 2006).

Portal do Biodiesel. 2006. http://www.biodieselbr.com/ (accessed February 2006).

Post, W.M. and K.C. Kwon. 2000. Soil carbon sequestration and land-use change: Processes and potential. *Global Change Biol.* 6: 317–327.

Powlson, D.S., A.B. Richie, and I. Shield. 2005. Biofuels and other approaches for decreasing fossil fuel emissions from agriculture. *Ann. Appl. Biol.* 146: 193–201.

Ragauskas, A.J. et al. 2006. The path forward for biofuels and biomaterials. *Science* 311: 484–489.

Ranga Rao, A. et al. 2007. Effect of salinity on growth of green alga *Botryococcus braunii* and its constituents. *Biores. Technol.* 98: 560–564.

Rede Baiana de Biocombustíveis. 2006. Informativo 119. www.redebaianadebiocombustiveis. ba.gov.br (accessed January 2009).

Reijnders, L. 2008. Do biofuels from microalgae beat biofuels from terrestrial plants? *Trends Biotechnol.* 26: 349–350.

Resck, D.V.S. et al. 2000. Impact of conversion of Brazilian cerrados to cropland and pasture-land on soil carbon pool and dynamics. In Lal, R. et al., Eds., *Global Climate Change and Tropical Ecosystems*. Boca Raton, FL: CRC Press, pp. 169–197.

Robertson, G.P., E.A. Paul, and R.R. Harwood. 2000. Greenhouse gases in intensive agri-culture: Contributions of individual gases to the radiative forcing of the atmosphere. *Science* 289: 1922–1925.

Saturnino, H.M. et al. 2005. Cultura do pinhão-manso (*Jatrofa curcas L.*): Belo Horizonte (MG). *Inf. Agropec.* 26: 44–78.

Sawayama, S., T. Minowa, and S.Y. Yokoyama. 1999. Possibility of renewable energy produc-tion and CO_2 mitigation by thermochemical liquefaction of microalgae. *Biom. Bioenergy* 17: 33–39.

Schuman, G.E., H.H. Janzen, and J.E. Herrick. 2002. Soil carbon dynamics and potential car-bon sequestration by rangelands. *Environ. Poll.* 116: 391–396.

Shapouri, H., J.A. Duffield, and M. Wang. 2002. The energy balance of corn ethanol: An update. Agricultural Economic Report 813. U.S. Department of Agriculture, Washington, D.C.

Sheehan, J. et al. 1998. Life cycle inventory of biodiesel and petroleum diesel for use in an urban bus. Report NREL/SR-580-24089. National Renewable Energy Laboratory, Golden, CO.

Sheehan, J. et al. 2004. *A Look Back at the U.S. Department of Energy's Aquatic Species Program: Biodiesel from Algae*. Report NREL/TP-580-24190. National Renewable Energy Laboratory, Golden, CO.

Shimizu, Y. 1996. Microalgal metabolites: A new perspective. *Annu. Rev. Microbiol.* 50: 431–465.

Shimizu, Y. 2005. Microalgal metabolites. *Curr. Opin. Microbiol.* 6: 236–243.

Silveira, A.M. et al. 2000. Simulation of the effects of land use changes in soil carbon dynamics in the Piracicaba river basin, São Paulo State, Brazil. *Pesq. Agropec. Bras.* 24: 389–399.

Six, J. et al. 2001. Source and composition of soil organic matter fractions between and within soil aggregates. *Eur. J. Soil Sci.* 52: 607–618.

Six, J. et al. 2004. The potential to mitigate global warming with no-tillage management is only realized when practised in the long term. *Global Change Biol.* 10: 155–160.

Smeets, E.M.W. et al. 2005. A bottom-up assessment and review of global bio-energy potentials to 2050: Progress in energy and combustion. *Science* 33: 56–106.

Soussana, J.F. et al. 2004. Carbon cycling and sequestration opportunities in temperate grasslands. *Soil Use Mgt.* 20: 219–230.

Spath, P.L. and M.K. Mann. 2004. Biomass power and conventional fossil systems with and without CO_2 sequestration. Report NREL/TP-510-32575. National Renewable Energy Laboratory, Golden, CO.

Spolaore, P. et al. 2006. Commercial applications of microalgae. *J. Biosci. Bioeng.* 101: 87–96.

Suarez, P.A.Z. and S.M.P. Maneghetti. 2007. 70^0 Aniversário do biodiesel em 2007: Evolução histórica e situação atual no Brasil. *Quím. Nova* 30: 2068–2071.

Superintendência Da Zona Franca De Manaus. 2003. Estudo de viabilidade econômica: Dendê. http://www.suframa.gov.br/ suframa_publicacoes.cfm (accessed June 2005).

Turner, B.T. et al. 2007. *Creating Markets for Green Biofuels.* Research Report UCB-ITSTSRC-RR- 2007–1. Institute of Transportation Studies, University of California at Berkeley. http://www.its.berkeley.edu/ publications/UCB/2007/TSRCRR/UCB-ITS-TSRC-RR-2007-1.pdf (accessed June 2007).

Updegraff, K., M.J. Baughman, and S.J. Taff. 2004. Environmental benefits of cropland conversion to hybrid poplar: Economic and policy considerations. *Biom. Bioenergy* 27: 411–428.

Van Gerpen, J. 1996. *Comparison of the Engine Performance and Emissions Characteristics of Vegetable Oil-Based and Animal Fat-Based Biodiesel.* Iowa State University Reports Database.

Wagner, L. 2007. *Biodiesel from Algae Oil*: Research Report. Mora Associates Ltd.

West, T. O., and G. Marland. 2002. A synthesis of carbon sequestration, carbon emissions, and net carbon flux in agriculture: Comparing tillage practices in the United States. *Agric. Ecosyst. Environ.* 91: 217–232.

Worldwatch Institute. 2006. Biofuels for transport: Global potential and implications for sustainable agriculture and energy for the 21st century. Washington, D.C. http://www.worldwatch.org/ system/files/EBF038.pdf (accessed March 2008).

5 Corn and Cellulosic Ethanol Problems and Soil Erosion

D. Pimentel

CONTENTS

INTRODUCTION

With shortages of fossil energy, especially oil and natural gas, the United States moved to convert corn grain into ethanol with the goal to make the nation oil-independent. Using more than 34% of all U.S. corn from 24 million acres in 2008 provided the nation with less than 1.7% of its oil consumption. A real tragedy! Because the corn ethanol project has become a disaster, interest in developing cellulosic ethanol is growing. Wood, grasses, and crop residues have also been proposed as ethanol fuels (Pimentel and Pimentel, 2008a). It may appear beneficial to use renewable plant materials for biofuels, but the use of crop residues and other biomasses to produce biofuels raises many concerns about major environmental problems (Pimentel, 2006).

Conflicts exist today in the use of land, water, energy, and other environmental resources for food and biofuel production. Although much land worldwide is occupied by grain and other crops, malnutrition remains the leading cause of death today (Pimentel et al., 2007). The World Health Organization reports that more than 3.7 billion people (56% of the global population) are currently malnourished (lacking protein, calories, iron, iodine, and vitamins) and that number increases steadily. Grains constitute more than 80% of the world food supply (via direct and indirect consumption). Unfortunately, the Food and Agricultural Organization of the United Nations reports that per capita grain production has declined for the past 23 years (FAO 1961–2006; Pimentel and Pimentel, 2008b). This suggests that the nutritional needs of the human population require increasing amounts of agricultural resources to produce food.

Food and biofuels depend on the same resources for production: land, water, and energy. In the United States, ~19% of all fossil energy is utilized in the food system: ~7% for agricultural production, ~7% for processing and packaging, and ~5% for preparation and distribution (Pimentel et al., 2007). The objective of this chapter is to analyze: (1) the reliance of corn and cellulosic ethanol for biofuels on the same land, water, and fossil energy as food production and (2) the characteristics of environmental impacts caused by soil erosion.

CORN ETHANOL

ENERGY USE IN CORN PRODUCTION

The conversion factor for converting corn into ethanol by fermentation in a large plant is about 1 liter of ethanol from 2.69 kg of corn grain (approximately 9.5 liters pure ethanol per bushel of corn) (Pimentel and Patzek, 2005). The production of corn in the United States requires significant energy and dollar investments for the 14 inputs such as labor, farm machinery, fertilizers, irrigation, pesticides, and electricity (Table 5.1). To produce an average corn yield of 9,400 kg/ha (149 bu/ac) of corn using up-to-date production technologies requires the expenditure of about 8.2 million kcal of energy inputs (mostly natural gas, coal, and oil) listed in Table 5.1. This is the equivalent of ~930 liters of oil equivalents (~25% of grain energy) expended per hectare of corn. The production costs total about $927/ha for 9,400 kg/ha or approximately 10 cents/kg ($2.54/bushel) of corn produced.

Full irrigation (in areas of insufficient or no rainfall) requires about 100 cm/ha of water over a growing season. Because only about 15% of U.S. corn production currently is irrigated (USDA, 1997a), only 8.1 cm/ha of irrigation was included for the growing season. On average, water for irrigation is pumped from a depth of 100 m (USDA, 1997a). On this basis, the average energy input associated with irrigation is 320,000 kcal/ha (Table 5.1).

ENERGY INPUTS FOR FERMENTATION AND DISTILLATION

The average costs (dollars and energy) for a large (245 to 285 million liters/year), modern dry grind ethanol plant are listed in Table 5.2. In the fermentation and

TABLE 5.1
Energy Inputs and Costs of Corn Production per Hectare in United States

Input	Quantity	kcal × 1,000	Cost ($)
Labor	11.4 hrs[a]	462[b]	148.20[c]
Machinery	55 kg[d]	1,018[e]	103.21[f]
Diesel	88 L[g]	1,003[h]	34.76
Gasoline	40 L[i]	405[j]	20.80
Nitrogen	155 kg[k]	2,480[m]	85.25[m]
Phosphorus	79 kg[n]	328[o]	48.98[p]
Potassium	84 kg[q]	274[r]	26.04[s]
Lime	1,120 kg[t]	315[u]	19.80
Seeds	21 kg[v]	520[w]	74.81[x]
Irrigation	8.1 cm[y]	320[z]	123.00[aa]
Herbicides	6.2 kg[bb]	620[ee]	124.00
Insecticides	2.8 kg[cc]	280[ee]	56.00
Electricity	13.2 kWh[dd]	34[ff]	0.92
Transport	204 kg[gg]	169[hh]	61.20
Totals		8,228	$926.97
Corn yield	9,400 kg/ha[ii]	33,840	kcal input:output 1:4.11

[a] NASS, 2003.
[b] It is assumed that a person works 2,000 hrs/yr and utilizes an average of 8,000 liters of oil equivalents per year.
[c] It is assumed that labor is paid $13 an hour.
[d] Pimentel and Pimentel, 1996.
[e] Prorated per hectare and 10-year life of machinery. Tractors weigh 6 to 7 tons and harvesters 8 to 10 tons; plows, sprayers, and other equipment also present.
[f] Hoffman et al., 1994.
[g] Wilcke and Chaplin, 2000.
[h] Input 11,400 kcal/liter.
[i] Estimated.
[j] Input 10,125 kcal/liter.
[k] NASS, 2003.
[l] Patzek, 2004.
[m] Cost 55 cents/kg.
[n] NASS, 2003.
[o] Input 4,154 kcal/kg.
[p] Cost 62 cents/kg.
[q] NASS, 2003.
[r] Input 3,260 kcal/kg.
[s] Cost 31 cents/kg.
[t] Brees, 2004.
[u] Input 281 kcal/kg.

TABLE 5.1 (CONTINUED)
Energy Inputs and Costs of Corn Production per Hectare in United States

[v] Pimentel and Pimentel, 1996.

[w] Pimentel and Pimentel, 1996.

[x] USDA, 1997b.

[y] USDA, 1997a.

[z] Batty and Keller, 1980.

[aa] Irrigation with 100 cm water/ha costs $1,000 (Larsen et al., 2002).

[bb] Larson and Cardwell, 1999.

[cc] USDA, 2002.

[dd] USDA, 1991.

[ee] Input 100,000 kcal/kg herbicide and insecticide.

[ff] Input 860 kcal/kWh requires 3 kWh thermal energy to produce 1 kWh electricity.

[gg] Goods transported include machinery, fuels, and seeds shipped ~1,000 km.

[hh] Input 0.83 kcal/kg/km transported.

[ii] Average. USDA, 2006; USCB, 2004–2005.

distillation process, the corn is finely ground and approximately 15 liters of water added per 2.69 kg of ground corn. Some of this water is recycled. After fermentation, obtaining 1 liter of 95% pure ethanol from the 8% to 12% ethanol beer and 92% to 88% water mixture, the single liter of ethanol must be extracted from approximately 11 liters of the ethanol/water mixture.

Although ethanol boils at about 78°C and water boils at 100°C, the ethanol is not extracted during the first distillation that yields 95% pure ethanol (Maiorella, 1985; Wereko-Brobby and Hagan, 1996; S. Lamberson, Cornell University, personal communication, 2000). For mixing with gasoline, the 95% ethanol must be further processed and more water removed, requiring additional fossil energy inputs to achieve 99.5% pure ethanol (Table 5.2). Thus, a total of about 10 liters of wastewater must be removed per liter of ethanol produced, and disposition of this relatively large amount of sewage effluent incurs energy, economic, and environmental costs.

Production of a liter of 99.5% ethanol uses 46% more fossil energy than the energy produced as ethanol and costs 45 cents per liter ($1.71 per gallon; Table 5.2). The corn feedstock requires about 32% of the total energy input. The total cost including energy inputs for fermentation and distillation and the apportioned energy costs of stainless steel tanks and other equipment totals $454.23 per 1,000 liters of ethanol produced (Table 5.2).

NET ENERGY YIELD

The largest energy inputs in corn ethanol production are required to produce the corn feedstock along with the steam energy and electricity used for fermentation and distillation. The total energy input to produce a liter of ethanol is 7,474 kcal (Table 5.2). However, a liter of ethanol has an energy value of only 5,130 kcal. Based

TABLE 5.2

Inputs per 1,000 Liters of 99.5% Ethanol Produced from Corn[a]

Input	Quantity	kcal × 1,000	Dollars
Corn grain	2,690 kg[b]	2,355[b]	265.27
Corn transport	2,690 kg[b]	322[c]	21.40[d]
Water	15,000 L[e]	90[f]	21.16[g]
Stainless steel	3 kg[i]	165[p]	10.60[d]
Steel	4 kg[i]	92[p]	10.60[d]
Cement	8 kg[i]	384[p]	10.60[d]
Steam	2,646,000 kcal[j]	2,646[j]	21.16[k]
Electricity	392 kWh[j]	1,011[j]	27.44[l]
95% ethanol to 99.5%	9 kcal/L[m]	9[m]	40.00
Sewage effluent	20 kg BOD[n]	69[h]	6.00
Distribution	331 kcal/L[q]	331	20.00[q]
Total		7,474	$454.23

[a] Output: 1 liter ethanol = 5,130 kcal (low heating value). Mean yield of 2.5 gal pure EtOH per bushel obtained from industry-reported ethanol sales less ethanol imports from Brazil, both multiplied by 0.95 to account for 5% by volume of #14 gasoline denaturant; result divided by industry-reported bushels of corn inputs to ethanol plants.

[b] http://petroleum.berkeley.edu/patzek/BiofuelQA/Materials/ TrueCostofEtOH.pdf; Patzek, 2006.

[c] Data from Table 5.1.

[d] Based on 144 km round trip.

[e] Pimentel, 2003.

[f] 15 liters of water mixed with each kilogram of grain.

[g] Pimentel et al., 2004.

[h] Pimentel et al., 2004.

[i] 4 kWh energy required to process 1 kg BOD (Blais et al., 1995).

[j] Estimated from industry reported costs of $85 million per 65 million gallons/yr dry grain plant amortized over 30 years. Total amortized cost is $43.6/1,000 liters EtOH, of which an estimated $32 go to steel and cement.

[k] Illinois Corn, 2004. Current estimate is below the average of 40,000 Btu/gal denatured ethanol paid to the Public Utilities Commission in South Dakota by ethanol plants in 2005.

[l] Calculated based on coal fuel; less than the 1.95 kWh/gal of denatured EtOH in South Dakota; see k.

[m] $.07 per kWh (USCB, 2004–2005).

[n] 95% ethanol converted to 99.5% for addition to gasoline (T. Patzek, University of California at Berkeley, personal communication, 2004).

[o] 20 kg BOD/1,000 liters of ethanol produced (Kuby et al., 1984).

[p] Newton, 2001.

[q] DOE, 2002.

on a net energy loss of 2,344 kcal of ethanol produced, 46% more fossil energy is expended than is produced as ethanol.

ECONOMIC COSTS

Current ethanol production technology uses more fossil fuel and incurs substantially higher costs than its energy value is worth on the market. Clearly, without the more than $6 billion federal and state government yearly subsidies, U.S. ethanol production would be reduced or cease, confirming that ethanol production is uneconomical (National Center for Policy Analysis, 2002). Federal and state subsidies for ethanol production of more than $6 billion/year are paid mainly to large corporations (Koplow, 2006), while corn farmers receive minimum profits per bushel for their corn (Pimentel and Patzek, 2005). Senator McCain reports that direct subsidies for ethanol plus subsidies for corn grain amount to 79 cents per liter (McCain, 2003).

If the production cost of a liter of ethanol were added to the tax subsidy cost, the total cost for a liter of ethanol would be $2.47. The mean wholesale price of ethanol was almost $1.00 per liter without subsidies. Because of its relatively low energy content, 1.6 liters of ethanol produce the energy equivalent of 1 liter of gasoline. Thus, the cost of producing an equivalent amount of ethanol to replace a liter of gasoline is $3.00 (or $11.34 per gallon of gasoline). This is far more than 53 cents, the current cost of producing a liter of gasoline. The subsidy per liter of ethanol is 60 times greater than the subsidy per liter of gasoline! This is why ethanol is so attractive to large corporations.

USE OF CORN LAND

Currently, about 34 billion liters of ethanol (9 billion gallons) are produced in the United States each year (EIA, 2009). The total petroleum fuel used in the U.S. was about 2,500 billion liters (USCB, 2008). Therefore, 34 billion liters of ethanol (energy equivalent to 20 billion liters of petroleum) provide only 1.7% of the petroleum utilized last year. To produce this 34 billion liters of ethanol, about 9.6 million ha or 34% of U.S. corn land was required. Expanding corn ethanol production to 100% of U.S. corn crops would fill only 5% of U.S. petroleum needs.

CELLULOSIC ETHANOL

Cellulosic ethanol is an imprecise term. It suggests that certain components of wood and green plant materials (cellulose, pectins, and hemicelluloses) can be chemically separated (from mostly lignin in wood and grasses) and partially split into hexose and pentose monomers that are then fermented to produce ethanol. Low energy industrial processes for producing ethanol from biomass do *not* exist. For that reason, not one single plant in the world produces ethanol from cellulosic biomass.

Cellulose is the principal structural component of cell walls in higher plants. It is the most abundant form of living terrestrial biomass (Pimentel, 2001). For hundreds of millions of years, cellulose has protected plants from the elements and animals, and from chemical attacks by fungi and bacteria. Cotton is 98% pure cellulose, flax is 80%, and wood is 40% to 50% cellulose; the remaining 50% to 60% consists of other complex polysaccharides (20% to 35% hemicelluloses and 15% to 35% lignin). The special properties of cellulose result from the association of the long, straight polymeric chains to form fibers called micro-fibrils that are stronger than steel. The micro-fibrils then form larger fibers in a criss-cross pattern and intermix with gel-like polysaccharides, hemicelluloses, and pectins that function as biocements (Taiz and Zeiger, 1998). The structure somewhat resembles fiberglass and other composite materials in which rigid crystalline fibers reinforce a more flexible matrix.

The β-glycosidic bonds are crucial in determining the structural properties of cellulose and thus the strength of its fibers. Because of the β bonds, the chain assumes an extended rigid configuration, with each glucose residue turned 180 degrees from its neighbor (Taiz and Zeiger, 1998). Another consequence of alternating top and bottom glucose residues is that OH groups of adjacent chains allow extensive hydrogen bonding. The extensive interchain hydrogen bonding and rigid β configuration make cellulose fibers very strong and able to resist severe environmental stresses including strong sodium hydroxide and acids. In summary, almost a billion years of plant evolution have made cellulose very stable and resistant to biochemical attacks. Cellulose can be decomposed quickly and hydrolyzed only by mechanical grinding, steam explosion, or severe chemical attack by hot concentrated sulfuric acid or sodium hydroxide. Biochemical enzymatic attacks take a long time and do not efficiently break down cellulose.

The process of separating cellulose fibers from the rest of woody biomass is well-known, fast, efficient, and very energy-intensive. Known as the paper kraft process, it is used to produce paper pulp and involves the use of caustic sodium hydroxide and sodium sulfide to extract the lignin from wood fiber in large pressure vessels called digesters. The process name is derived from *kraft*, a German word meaning *strong*. Unfortunately, the best energy efficiency is ~6,200 kcal/kg of paper—more than the high heating value of pure ethanol (5,130 kcal). Therefore a much milder, enzymatic process must be used to obtain simple sugars from cellulose.

OBSTACLES TO CELLULOSIC ETHANOL PRODUCTION

BIOMASS AVAILABILITY

Using food crops such as corn grain to produce ethanol raises major nutritional and ethical concerns. Nearly 60% of humans in the world are currently malnourished, and the need for grains and other basic foods is critical (WHO, 2005). Growing crops for fuel squanders land, water, and energy resources vital to produce food. Using corn for ethanol increases the price of U.S. beef, chicken, pork, eggs, breads, cereals, and milk by 50, to 80%. In addition, Jacques Diouf, director general of the United Nations Food and Agriculture Organization, reports that using food grains to produce biofuels has already caused food shortages for the poor of the world (Mailstrom, 2008). Growing crops for biofuel ignores the need to reduce natural resource consumption and exacerbates malnourishment worldwide by turning precious grains into biofuels. The U.S. Congressional Budget Office reports that Americans are paying $6 to $9 billion more per year in grocery bills (USCB, 2007).

TABLE 5.3

Total Biomass and Solar Energy Captured Annually in United States

Crops	901×10^6 t	14.4×10^{15} BTU
Pastures	600×10^6 t	9.6×10^{15} BTU
Forests	527×10^6 t	8.4×10^{15} BTU
Total	$2,028 \times 10^6$ t	32.4×10^{15} BTU

Note: An estimated 32×10^{18} BTU of sunlight reaching the U.S. per year suggests that green plants in the U.S. collect 0.1% of solar energy. (*Sources*: Adapted from Jölli and Giljum, 2005; Crop Harvest, 2007; Crop Production, 2007; Forest Service, 2007.)

Recent policy decisions have mandated increased production of biofuels in the United States and elsewhere. In the Energy Independence and Security Act of 2007, President Bush set "a mandatory renewable fuel standard (RFS) requiring fuel producers to use at least 36 billion gallons of biofuel in 2022." This would require 1.6 billion tons of biomass harvested per year, or 80% of all biomass in the U.S. from agricultural crops, grasses, and forests (Table 5.3). This huge biomass harvest would decimate biodiversity and food supplies in the U.S.!

Biomass Availability

Natural *net* productivity of a mature ecosystem (an earth household, e.g., a forest or grassland) is low on a human time scale. (Very slow carbon burial occurs on a geological time scale.) For example, oil, natural gas, and coal supplies have accumulated for more than 700 million years. What is produced by autotrophic plants and rock weathering is consumed by heterotrophs (bacteria, fungi, and other organisms that are continuously recycled as nutrients for plants). Some bacteria and fungi, in return for food from plant roots, capture nitrogen from the air and convert it to ammonia, thus providing natural fertilization. *Biowaste* is an engineering classification of plant and animal parts not consumed in an industrial process. This dated human concept is completely alien to natural ecosystems that must recycle their matter completely to survive. Excessive biowaste removal robs ecosystems of vital nutrients and species and degrades them irreversibly. As discussed by Patzek and Pimentel (2005), ecosystems from which we remove biomass at high rate (crop fields, tree plantations) must be heavily subsidized with fossil energy and earth minerals.

Contamination of Cellulosic Feedstock

For corn starch fuel ethanol, normal fermentation time in batch mode (no continuous reactors in operation) is 48 hours; up to 72 hours is acceptable. These estimates do not include downtime, cleaning, start-up, and other management needs. The number of failures increases exponentially beyond 72 hours due to contamination with aceto-gens and other bacteria. As noted in the literature, typical enzyme processes for lingo-cellulosic alcohol take 5 to 7 days (120 to 170 hr). The long enzymatic step creates serious problems if lingo-cellulosic ethanol producers scale up production beyond laboratory or pilot scale to a conventional fermentation vessel that *cannot* be sterilized for 120 to 170 hours.

Enzyme Yield versus Rate

The rate of ligno-cellulose hydrolysis and fermentation can be increased by pre-treatment (such as ball milling to exceedingly fine dust, at enormous energy cost, or steam exploding with acid), but rates will slow rapidly before high yields are obtained. The main problem is the number of binding sites available—the outside-in rate limitation phenomenon. It simply takes time to chew up sturdy lingo-cellulosic particles. Of course, one could run the lingo-cellulose through a kraft-like process, but this cannot be done for lingo-cellulosic ethanol because of enormous energy losses. One can obtain acceptable yields and rates by performing energy-intensive (and unaffordable) pretreatment or (relatively) high yields with modest pre-treatment and a long wait (ideally for weeks). Thus, despite claims to the contrary, an effective industrial technique for producing lingo-cellulosic ethanol does not yet exist; any technique may never achieve a sufficiently favorable energy balance.

Thermodynamics

Current energy efficiency of producing cellulosic ethanol is so low that all other investigated paths to liquid biofuels like corn ethanol are better (Patzek and Pimentel, 2005). For this reason, not a single commercial cellulosic ethanol plant exists. The average energy input per hectare for switchgrass production is only about 3.9 million kcal/year (Table 5.4). An exceptional average yield of 10 t/ha/yr suggests for each kcal invested as fossil energy the return is 11 kcal—an excellent return. However, this return is *impossible* to realize longer than a year in any environment other than an ecologically balanced prairie.

Nonetheless, massive industrial *monocultures* of switchgrass have been proposed and studied. If pelletized for use as a fuel in stoves, the return is reported to be about 14.6 kcal (Samson et al., 2004). The 14.6 is higher than the 11 kcal cited in Table 5.5, because a few more inputs were included than in the Samson report. If the realistic sustained yield of switchgrass were 1 to 4 t/ha/yr, the 14.6 return would drop to 1.5 to 4, similar to corn return. The cost per ton of switchgrass pellets ranges from $94 to $130 (Samson et al., 2004)—ostensibly an excellent price per ton, but converting switchgrass into ethanol yields a negative energy return (Table 5.5). The energy required to produce 1 liter of ethanol using switchgrass is 1.81 liters of oil equivalents—significantly more negative than producing corn ethanol (Pimentel and Patzek, 2008). The cost of producing a liter of ethanol using switchgrass is $1.75 (Table 5.5); the two major energy inputs for switchgrass conversion into ethanol were steam and electricity.

TABLE 5.4

Average Annual Input and Energy Inputs per Hectare for Switchgrass Production

Input	Quantity	10^3 kcal	Dollars
Labor	5 hr[a]	200[b]	65[c]
Machinery	30 kg[d]	555	50[a]
Diesel	150 L[e]	1,500	160
Nitrogen	80 kg[e]	1,280	90[e]
Seeds	1.6 kg[f]	100[a]	3[f]
Herbicides	3 kg[g]	300[h]	30[a]
Total	10,000 kg yield[i]	3,935	$398[j]
	40 million kcal yield	input/output ratio = 1:02[k]	

[a] Estimated.

[b] Average person works 2,000 hours/yr and uses about 8,000 liters of oil equivalents.

[c] Prorated = 200,000 kcal.

[d] Agricultural labor is paid $13/hr.

[e] Machinery estimate also includes additional 25% for repairs.

[f] Calculations based on current data.

[g] Samson, 1991.

[h] Calculations based on Henning, 1993.

[i] 100,000 kcal/kg of herbicide.

[j] Samson et al., 2000.

[k] Brummer et al., 2000 estimated cost of about $400/ha for switchgrass produc-tion. Thus, the $268 total is about 49% lower than Brummer et al. estimated and includes several inputs not included by Brummer et al. Samson et al., 2000 estimated input:output return of 1:14.9; we added several inputs they did not include. Input:output return of 1:11 would be excellent if sustained yield of 10 t/ha/yr were possible.

SOIL EROSION AND DEGRADATION ASSOCIATED WITH BIOFUELS

Soil erosion and land degradation particularly concern agriculturalists and foresters because of increasing biofuel production, especially cellulosic ethanol. Serious soil erosion in U.S. agricultural systems continues, with estimated soil losses 11 times the sustainable rate (Peak Soil, 2008). The prime cause of this high erosion is the deple-tion of cellulosic biomass cover that protects soil from rainfall and wind energy. Row crops such as corn and soybeans are particularly susceptible to erosion (Pimentel, 2006). Tilling for planting of row crops leaves the soil vulnerable to wind and rain-fall. After harvesting, soybeans yield little crop residue that covers only 20% of the crop land. Corn stover covers about 60% of land after harvest (Leiting, 2007).

Most corn ethanol is produced from corn grown continuously because continuous corn crops provide the most profit for farmers. The soil erosion rate for corn grown continuously on the same land is at least 15 t/ha/yr; thus, soil is lost 15 times faster

TABLE 5.5
Inputs per 1000 Liters of 99.5% Ethanol Produced from U.S. Switchgrass[a]

Input	Quantity	kcal × 1000	Dollars
Switchgrass	9,600 kg[b]	3,778	960
Transport	9,600 kg[b]	600[c]	90[d]
Water	250,000 L[e]	140[f]	40[m]
Stainless steel	3 kg[g]	165[g]	11[g]
Steel	4 kg[g]	92[g]	11[g]
Cement	8 kg[g]	384[g]	11[g]
Grinding	9,600 kg	400[h]	32[h]
Sulfuric acid	460 kg[i]	0	323[n]
Steam	15.6 t[i]	8.482	69
Lignin	2,400 kg[j]	−2,880	−23
Electricity	999 kWh[i]	2,555	88
Ethanol: 95 to 99.5%	13.5 kcal/L[k]	13.5	77
Sewage effluent	60 kg BOD[l]	207[o]	23
Distribution	497 kcal/L[p]	497	29
Total		14,434	$1,754

[a] Output: 1 liter ethanol = 5,130 kcal. Ethanol yield = 1,000 L/9,600 kg dry biomass (dbm); estimate appears sound (Saeman, 1945; Harris et al., 1945; Conner et al., 1986).

[b] Data from Table 5.4.

[c] Based on 144 km round trip.

[d] Pimentel, 2003.

[e] Pimentel, 2003.

[f] 15 liters of water mixed with each kilogram of biomass.

[g] Newton, 2001.

[h] Wood Tub Grinders, 2004.

[i] Estimate based on Arkenol, 2004.

[j] Wood is about 25% lignin; removing most of the water by filtering can reduce moisture level 200% (Crisp, 1999).

[k] 95% ethanol converted to 99.5% for addition to gasoline (T. Patzek, University of California at Berkeley, personal communication, 2004).

[l] 20 kg of BOD/1,000 liters ethanol produced (Kuby et al., 1984).

[m] Pimentel, 2003.

[n] Sulfuric acid sells for $7/kg.

[o] 4 kWh of energy required to process 1 kg BOD (Blais et al., 1995).

[p] DOE, 2002.

than soil is formed (Pimentel, 2006). SOM in these soils averages only about 2% (Lickacz and Penny, 2001). In organic corn production (corn grown in rotation with soybeans plus a vetch or a similar cover crop), the SOM ranged from 5.3% to 5.5% (Pimentel et al., 2005). Soil loss under these soil conservation conditions was less than 1 t/ha/yr—less than soil loss via conventional corn production.

SOM provides a base for productive organic farming and sustainable agriculture. The SOM in the upper 15 cm of soil in the Rodale organic farming systems was approximately 110,000 kg/ha (Pimentel et al., 2005). About 41% of the volume of organic matter in organic systems consisted of water, compared to only 35% in conventional systems (Sullivan, 2002). The amount of water held in organic soils was estimated to be 816,000 liters/ha. This large amount of stored water made organic farming systems more tolerant to drought (Pimentel et al., 2005).

Large amounts of SOM increase soil biodiversity (Lavelle and Spain, 2001; Pimentel et al., 2006). The arthropod numbers range from 2 million to 5 million/ha; earthworms, from 1 million to 5 million/ha (Lavelle and Spain, 2001; Gray, 2003), producing as many as 1,000 earthworm and insect holes per square meter. These large holes can be particularly helpful in increasing the percolation of water into the soil and reducing water runoff. Earthworms and insects increase soil quality and microbes can be especially beneficial. Soil bacteria and fungi together can weigh about 6,000 kg/ha (Pimentel et al., 2006) and more than 20,000 species of these organisms can be present in soil (Pimentel et al., 2006).

The intensive application of nitrogen fertilizers in corn production was perceived to sequester soil organic carbon in the soil. However, after a 40- to 50-year application of synthetic nitrogen fertilizer in Illinois, a net decline of soil carbon occurred despite the massive residue incorporation (Khan et al., 2007).

Some investigators (Tilman et al., 2006; Perlack et al., 2005) suggest that crop residues can be harvested for biofuel production. However, the removal of residues will increase soil erosion from 10- to 100-fold (Pimentel and Lal, 2007; Lal and Pimentel, 2007). *Removal of crop residues will devastate U.S. agriculture.*

Close-grown crops, like wheat, that produce an average 7 t/ha/yr, protect the soil from erosion by maintaining an erosion rate of about 5 t/ha/yr. This is better than the rate for row crops like corn that produce an average soil erosion rate of 15 t/ha/yr (Troeh et al., 2004). After germinating, spring wheat grows fast and develops a relatively dense stand of vegetation cover of 150 to 200 plants/m^2. This density is capable of protecting the soil from rain and wind energy, but wheat is not a good biofuel crop because of its low yield (USDA, 2006).

Some crops, such as grass, provide nearly complete soil cover after they are established. These crops are usually grown as perennials and cover the soil all year for about five years. The soil erosion rate from continuous grass is reported as only 0.1 to 1 t/ha/yr (Lal, 2004), but grass and similar crops are not generally productive as biofuel crops. Sugar cane, a common biofuel crop, exhibits high soil erosion rates reported around 148 t/ha/yr in Australia (Prove et al., 1986). No-till and ridge till planting of corn and similar crops will reduce soil erosion from 18 to 2 t/ha/yr (Wortmann and Jasa, 2003). Herbicides and other pesticides are needed for no-till corn production, but herbicides may not be necessary for ridge till crops.

The water-holding capacity and nutrient levels of soil decline when erosion occurs. With conventional corn production, erosion reduced the volume of moisture in the soil about 50% compared with organic corn (Pimentel et al., 2005). When conservation technologies like organic agriculture are employed, yields may increase because water, nutrients, and SOM are retained. For example, in Pennsylvania, corn and soybean yields were 33 to 50% higher in organic systems when SOM increased over time even in drought conditions (Pimentel et al., 2005).

Undisturbed forests often have dense soil covers of leaves, twigs, and other organic matter; these forest ecosystems have soil erosion rates typically ranging from <0.1 to 0.2 t/ha/yr (Pimentel, 2006). The combination of organic mulch, tree cover, and tree roots makes most natural forest soils, even on steep slopes of 70%, resistant to erosion and rapid water runoff. Forests lose significant quantities of water, soil, and nutrients when cut and harvested (Mongabay, 2004) and erosion rates increase. Therefore, the use of forests for producing biofuels will increase rates of soil erosion. However, short-rotation woody crops have been shown to improve groundwater quality and water runoff in comparison to row crops (Thornton et al., 1996).

CONCLUSIONS

Rapidly growing world population and rising consumption of fossil fuels have increased demands for both food and biofuels and ultimately exaggerate shortages of both commodities. Producing corn and cellulosic biofuels requires huge amounts of fossil energy, water, and land resources and will thus intensify conflicts for these resources. Using food and cellulosic biomass crops to produce ethanol raises major nutritional and ethical concerns. The malnutrition already common in certain human populations clearly shows that needs for grains and basic foods are critical.

Mandated policies such as the U.S. Energy Independence and Security Act of 2007 that calls for a "mandatory renewable fuel standard (RFS) requiring fuel producers to use at least 36 billion gallons of biofuel in 2022" would mean the harvest of 1.6 billion tons of biomass per year, or 80% of all biomass in the U.S. including all agricultural crops, grasses, and forests. Increased biofuel production will impact the quality of food and cellulosic plants in cropping systems. The release of large quantities of carbon dioxide associated with the planting and processing of plant materials for biofuels is reported to reduce the nutritional qualities of major foods including wheat, rice, barley, potatoes, and soybeans (Southwestern University, 2008). Protein levels in crops grown at high levels of carbon dioxide may be reduced as much as 15%.

The biofuels created to diminish dependence on fossil fuels are *increasing* our dependence on fossil fuels. In most cases, biofuel production requires more fossil energy than the quantity of energy produced (Tables 5.2 and 5.5). Examples of negative energy returns are corn ethanol, −48%; switchgrass, −50%; soybean biodiesel, −53%; and rapeseed, −58% (Pimentel et al., 2008). The U.S. imports oil and natural gas to produce biofuels, thus increasing its oil dependence.

Publications promoting biofuels have based their claims on incomplete or insufficient data. For example, claims that cellulosic ethanol provides net energy (Tilman et al., 2006) have not been experimentally verified; most of the calculations are theoretical. Finally, environmental problems including water pollution from fertilizers and pesticides, global warming, and air pollution intensify with biofuel production. The world simply does not have sufficient land, water, and energy available to produce biofuels. Increased use of biofuels erodes soil, reduces available food crop land, and threatens the world food system.

REFERENCES

Arkenol. 2004. Our technology: Concentrated acid hydrolysis. www.arkenol.com/Arkenol%20 Inc/tech01.html (accessed Aug. 2, 2004).

Batty, J.C. and J. Keller. 1980. Energy requirements for irrigation. In *Handbook of Energy Utilization in Agriculture,* Pimentel, D., Ed., CRC Press, Boca Raton, FL, pp. 35–44.

Blais, J.F. et al. 1995. Les esures deficacite energetique dans le secteur de leau. J.L. Sassville and J.F. Balis, Eds., Les Mesures deficacite Energetique pour Lepuration des eaux Usees Municipales. Scientific Report 405. Vol. 3. INRS-Eau, Quebec.

Brees, M. 2004. Corn silage budgets for Northern, Central and Southwest Missouri. http:// www.agebb.missouri.edu/mgt/budget/fbm-0201.pdf (accessed Sept. 1, 2004).

Brown, L.R. 2008. Why ethanol production will drive world food prices even higher. Earth Policy Institute, http://www.earthpolicy.org/Updates/2008/Update69.htm. (accessed Jan. 24, 2008).

Brummer, E.C. et al. 2000. *Switchgrass Production in Iowa: Economic Analysis, Soil Suitability, and Varietal Performance.* Iowa State University, Ames. www.iowaswitchgrass.com/__ docs/pdf/Switchgrass%20in%20Iowa%202000.pdf (accessed Oct. 6, 2005).

Conner, A.H. and L.F. Lorenz. 1986. Kinetic modeling of hardwood prehydrolysis III: Water dilute acetic acid prehydrolysis of southern red oak. *Wood Fiber Sci.* 18: 248–263.

Crisp, A. 1999. Wood residue as an energy source for the forest products industry. Australian National University. http://sres.anu.edu.au/associated/fpt/nwfp/woodres/woodres.html (accessed July 10, 2006).

Crop Harvest. Biological System Engineering: Crop Systems, Washington State University, 2007. http://www.bsyse.wsu.edu/cropsyst/manual/simulation/crop/harvest.htm (accessed Jan. 29, 2008).

Crop Production. 2007. National Agricultural Statistical Services, U.S. Department of Agriculture. http://www.usda.gov/nass/PUBS/TODAYRPT/crop0907.txt (accessed Jan. 29, 2008).

DOE. 2002. Review of transport issues and comparison of infrastructure costs for a renewable fuel standard. U.S. Department of Energy. http://tonto.eia.doe.gov/FTPROOT/service/ question3.pdf (accessed Oct. 8, 2002).

Energy Information Administration. 2009. Official energy statistics from the U.S. Government. http://tonto.eia.doe.gov/dnqv/pet/pet_som_Snd_d_nus_mbbl_m_cur.htm.

FAO. 1961–2006. Food Balance Sheets. Food and Agriculture Organization, United Nations, Rome.

Forest Service. 2007. Forest Inventory and Analysis RPA Assessment Tables. U.S. Department of Agriculture.

Gray, M. 2003. Influence of agricultural practices on earthworm populations. *Pest Mgt. Crop Dev. Bull.* www.ag.uiuc.edu/cespubs/pest/articles/200305d.html (accessed April 26, 2005).

Jölli, D. and S. Giljum. 2005. Unused biomass extraction in agriculture, forestry and fisheries, SERI Studies 3, Sustainable Europe Research Institute, Vienna.

Harris, E.E. et al. 1945. Hydrolysis of wood: Treatment with sulfuric acid in a stationary digester. *Ind. Eng. Chem.* 37: 12–23.

Henning, J.C. 1993. Big Bluestem, Indiangrass and Switchgrass. Department of Agronomy, University of Missouri, Columbia.

Hoffman, T.R., W.D. Warnock and H.R. Hinman. 1994. Crop Enterprise Budgets, Timothy-Legume and Alfalfa Hay, Sudan Grass, Sweet Corn and Spring Wheat under Rill Irrigation, Kittitas County, Washington. Farm Business Report EB 1173. Washington State University.

Illinois Corn. 2004. Ethanol's Energy Balance. http://www.ilcorn.org/Ethanol/Ethan_Studies/ Ethan_Energy_Bal/ethan_energy_bal.html (accessed Aug. 10, 2004).

Khan, S.A. et al. 2007. The myth of nitrogen fertilization for soil carbon sequestration. *J. Environ. Qual.* 36: 1821–1832.

Koplow, D. 2006. Biofuels at What Cost? Government Support for Ethanol and Biodiesel in the United States. Global Studies Initiative of International Institute for Sustainable Development. http://www.globalsubsidies.org/IMG/pdf/biofuels_subsidies_us.pdf (accessed Feb. 16, 2007).

Kuby, W.R., R. Markoja, and S. Nackford. 1984. *Testing and Evaluation of On-Farm Alcohol Production Facilities: Acures Corporation.* U.S. Environmental Protection Agency: Cincinnati.

Lal, R. 2004. Soil carbon sequestration impacts on global climate change and food security, *Science* 34: 1623–1627.

Lal, R. and D. Pimentel. 2007. Biofuels from crop residues. *Soil Tillage Res.* 93: 237–238.

Larsen, K., D. Thompson and A. Harn. 2002. Limited and Full Irrigation Comparison for Corn and Grain Sorghum. http://www.colostate.edu/Depts/SoilCrop/extension/Newsletters/2003/Drought/sorghum.html (accessed Sept. 2, 2002).

Larson, W.E. and V.B. Cardwell. 1999. History of U.S. Corn Production. http://citv.unl.edu/cornpro/html/history/history.html (accessed Sept. 2, 2002).

Lavelle, P. and A.V. Spain. 2001. *Soil Ecology.* Kluwer: Dordrecht.

Leiting, K.B. 2007. Agronomy: Corn and Soybean Crop Residue Management. Natural Resource Conservation Service. Albuquerque, NM. http://www.nm.nrcs.usda.gov/technical/tech-notes/agro/ag67.pdf (accessed Oct. 9, 2007).

Lickacz, J. and D. Penny. 2001. Plant Industry Division, Agriculture and Rural Development, Alberta Government. http://www1.agric.gov.ab.ca/$department/deptdocs.nsf/all/agdex890 (accessed Sept. 23, 2008).

Mailstrom. 2008. Food inflation and food shortages. http://mailstrom.blogspot.com/search?q=food+inflation+ (accessed Jan. 22, 2008).

Maiorella, B. 1985. Ethanol. In *Comprehensive Biotechnology,* Vol. 3, Blanch, H.W. et al., Eds. Pergamon, New York, Chap. 43.

McCain, J. 2003. Statement of Senator McCain on the Energy Bill. Press Release. November.

Mongabay. 2004. Forest Erosion. http://www.mongabay.co/0903.htm (accessed Dec. 7, 2007).

NASS. 2003. National Agricultural Statistics Service. http://usda.mannlib.cornell.edu (accessed Nov. 5, 2004).

National Center for Policy Analysis. 2002. Ethanol Subsidies, National Center for Policy Analysis. http://www.ncpa.org/pd/ag/ag6.html (accessed Sept. 9, 2002).

Newton, P.W. 2001. Human Settlements. Australian State of the Environment Report. http://www.deh.gov.au/soe/2001/settlements/acknowledgement.html (accessed Oct. 6, 2005).

Patzek, T.W. 2004. Thermodynamics of the corn–ethanol biofuel cycle. *Crit. Rev. Plant Sci.* 23: 519–567.

Patzek, T.W. and D. Pimentel. 2005. Thermodynamics of energy production from biomass. *Crit. Rev. Plant Sci.* 24: 327–364.

Peak Soil. 2008. Why biofuels are not sustainable and a threat to America's national security. http://www.ea2020.org/drupal/node/39 (accessed Jan. 31, 2008).

Perlack, R.D. et al. 2005. Biomass as Feedstocks for a Bioenergy and Bioproducts Industry: The Technical Feasibility of A Billion-Ton Annual Supply. Report ORNL/TM-2005/66. Oak Ridge National Laboratory.

Pimentel, D. 2001. Limitations of biomass energy. In *Encyclopedia of Physical Science and Technology.* 3rd ed., Meyers, R., Ed., Academic Press, San Diego, pp. 159–171.

Pimentel, D. 2003. Ethanol fuels: Energy balance, economics, and environmental impacts are negative. *Nat. Resources Res.* 12: 127–134.

Pimentel, D. 2006. Soil erosion: A food and environmental threat. *Environ. Dev. Sustain.* 8: 119–137.

Pimentel, D. and R. Lal. 2007. Biofuels from crop residues. *Soil Tillage Res.* 93: 237–238.

Pimentel, D. and T. Patzek. 2005. Ethanol production using corn, switchgrass, and wood: Biodiesel production using soybean and sunflower. *Nat. Resources Res.* 14: 65–76.

Pimentel, D. and T. Patzek. 2008. Ethanol production using corn, switchgrass and wood: Biodiesel production using soybean. In *Biofuels, Solar and Wind as Renewable Energy Systems,* Pimentel, D., Ed., Springer, Dordrecht, pp. 375–396.

Pimentel, D. and M. Pimentel. 1996. *Food, Energy and Society,* University of Colorado Press, Boulder.

Pimentel, D. and M. Pimentel. 2008a. *Food, Energy and Society.* 3rd ed. CRC Press, Boca Raton, FL.

Pimentel, D. and M. Pimentel. 2008b. Ecological engineering: Human population growth. In *Encyclopedia of Ecology,* Jorgensen, S.E. and B.D. Fath, Eds., Elsevier, Oxford, pp. 1907–1912.

Pimentel, D. et al. 2004. Water resources: Current and future issues. *BioScience* 54: 909–918.

Pimentel, D., P. Hepperly, J. Hanson et al. 2005. Environmental, energetic, and economic comparisons of organic and conventional farming systems. *Bioscience* 55: 573–582.

Pimentel, D., T. Petrova, M. Riley et al. 2006. Conservation of biological diversity in agricultural, forestry, and marine systems. In *Focus on Ecology Research,* Burk, A.R., Ed., Nova, New York, pp. 151–173.

Pimentel, D., S. Cooperstein, H. Randell et al. 2007. Ecology of increasing diseases: Population growth and environmental degradation. *Hum. Ecol.* 35: 653–668.

Prove, B.G., P.N. Truong, and D.S. Evans. 1986. Strategies for controlling cane land erosion in the wet tropical coast of Queensland. Proceedings of Australian Society Sugar Cane Technology Conference, pp. 77–84.

Samson, R. 1991. *Switchgrass: A Living Solar Battery for the Prairies.* Ecological Agriculture Projects, McGill University.

Samson, R., P. Duxbury, M. Drisdale et al. 2000. Assessment of Pelletized Biofuels. PERD Program, Natural Resources Canada.

Samson, R., P. Duxbury, and L. Mulkins. 2004. Research and Development of Fibre Crops in Cool Season Regions of Canada. Resource Efficient Agricultural Production Canada. http://www.reap-canada.com/Reports/italy.html (accessed June 26, 2004).

Saeman, J.F. 1945. Kinetics of wood saccharification: Hydrolysis of cellulose and decomposition of sugars in dilute acid at high temperature. *Ind. Eng. Chem.* 37: 43–52.

Southwestern University. 2008. Biology research finds rising CO_2 levels could decrease the nutritional value of major food crops. http://www.blackwell-synergy.com/doi/full10.1111/j.1365-2486.2007.01511.x. (accessed Feb. 15, 2008).

Sullivan, P. 2002. Drought Resistant Soil, Fayetteville (AR): Appropriate Technology Transfer for Rural Areas. www.attra.org/attra-pub/PDF/drought.pdf (accessed April 22, 2005).

Taiz, L. and E. Zeiger. 1998. *Plant Physiology.* Sinauer, Sunderland, MA.

Thornton, F.C. et al. 1996. Environmental impacts of converting cropland to short-rotation woody crop production: First year results. In *Proceedings of Seventh National Bioenergy Conference,* Nashville, TN, September, pp. 210–216.

Tilman, D., J. Hill, and C. Lehman. 2006. Carbon-negative biofuels from low-input high-diversity grassland biomass. *Science* 314: 1598–1600.

Troeh, F.R., A.H. Hobbs, and R.L. Donahue. 2004. *Soil and Water Conservation.* Prentice Hall, New York.

USCB. 2004–2005. *Statistical Abstract of the United States.* U.S. Government Printing Office, Washington, DC.

USCB. 2007. *Statistical Abstract of the United States: 2007.* Government Printing Office, Washington, DC.

USDA. 1991. Corn State: Costs of Production. U.S. Department of Agriculture, Washington.

USDA. 1997a. Farm and Ranch Irrigation Survey: 1997 Census of Agriculture. Vol. 3, Special Studies, Washington.

USDA. 1997b. 1997 Census of Agriculture. U.S. Department of Agriculture. Washington. http://www.ncfap.org (accessed Aug. 28, 2002).

USDA. 2002. *Agricultural Statistics*. U.S. Department of Agriculture, Washington.

USDA. 2006. *Agricultural Statistics*. U.S. Department of Agriculture. Washington.

Wereko-Brobby, C. and E. B. Hagan. 1996. *Biomass Conversion and Technology.* John Wiley & Sons, Chichester.

WHO. 2005. Malnutrition worldwide. World Health Organization. http://www.mikeschoice. com/reports/malnutrition_worldwide.htm (accessed Dec. 7, 2007).

Wilcke, B. and J. Chaplin. 2000. Fuel saving ideas for farmers. Minnesota/Wisconsin Engineering Notes. http://www.bae.umn.edu/extens/ennotes/enspr00/fuelsaving.htm (accessed Sept. 2, 2004).

Wood Tub Grinders. 2004. http://p2library.nfesc.navy.mil/P2_Opportunity_Handbook/7_ III_13.html (accessed Aug. 3, 2004).

Wortmann, C.S. and P.J. Jasa. 2003. Choosing the Right Tillage System for Row Crop Production. Institute of Agriculture and Natural Resources, University of Nebraska, Lincoln.

6 Ethanol Production from Sugarcane and Soil Quality

M.V. Galdos, Carlos Clemente Cerri,
M. Bernoux, and Carlos Eduardo P. Cerri

CONTENTS

INTRODUCTION

Sugarcane (*Saccharum officinarum* L.), a crop originally from New Guinea, was one of the first tropical crops to be adapted to large-scale farming (Brandes, 1956). It is a C4 plant—highly efficient in turning solar radiation into biomass. Sugarcane is a perennial crop harvested approximately annually, with up to six cycles before replanting. Sugarcane is produced commercially in more than 70 countries with nearly 22 million ha harvested annually, mostly between latitudes of 35°N to 35°S. The main producers are Brazil, India, and China, accounting for 33%, 23%, and 7% of 2007 total production, respectively (FAOSTAT, 2009).

Bioethanol can be produced from several sources of biomass including corn (*Zea mays*), beets (*Beta vulgaris* L.), wood residues, and sugarcane. Due to its vigorous growth, photosynthetic efficiency, and a production system that often includes the use of crop residues to generate power for processing mills, sugarcane is one of the

137

most attractive feedstocks for bioethanol production based on energy input-to-output ratios and greenhouse gas emission reductions compared to fossil fuels (Marris, 2006; Goldemberg, 2007).

Areas currently planted with sugarcane are undergoing significant expansion due to growing demands for bioethanol, driven by environmental, geopolitical, and economic issues. It is important to understand the implications of sugarcane growing on soil quality based on the vital role soil plays in sustainability of agricultural activities. The present chapter will briefly address the main impacts of sugarcane management practices on soil quality.

SUGARCANE MANAGEMENT PRACTICES AND SOIL QUALITY

HARVEST

Preharvest Burning

In the sugarcane production system, residue burning is a common practice to facilitate more efficient manual harvest and transport operations. After burning, the partially burned tops are separated from the stalks and left on the field. The burning of sugarcane before harvest represents 11% of the harvested residues burned annually (IPCC, 1995). The above-ground biomass is composed of approximately 60% to 80% stalks; the remainder consists of leaves and tops. According to field experiments with controlled preharvest fires, 70% to 95% of the dry matter of leaves and tops is lost with preharvest burning (Mitchel et al., 2000). Part of the charcoal produced during the burning events is carried by air currents, reaching surrounding areas. Besides the nuisance of charcoal deposition in residential areas, evidence indicates a link between sugarcane residue burning and respiratory diseases affecting populations in areas near production fields.

The release of respirable (<4 μm) silica minerals found in ash and aerosol produced by burning sugarcane can cause respiratory diseases (Le Blonde et al., 2008). Besides direct human health effects, environmental impacts arise from burning these residues. High concentrations of dissolved organic carbon in the atmosphere have been measured in sugarcane production regions (Coelho et al., 2008). Crutzen and Andreae (1990) linked biomass burning to regional production of O_3 and photochemical smog, increased acid deposition, and potential loss of fixed nitrogen (pyrodenitrification). Although the emitted CO_2 is reabsorbed by the next crop through photosynthesis, that does not occur with CH_4 and N_2O, contributing to the greenhouse effect—an argument against the global warming mitigation potential of biofuels (Scharlemann and Laurance, 2008).

Unburned Sugarcane Harvest

For environmental, agronomic, and economical reasons, the manual harvest of sugarcane with burning (Figure 6.1a) has been replaced by mechanical methods with retention of dry leaves and tops (trash) in the fields (Figure 6.1b) in a system called green cane management. The process includes the deposition of large amounts of plant litter on the soil (Figure 6.2) after each harvest, ranging from 10 to 20 tons of dry matter per hectare. The mulch formed impacts the whole production process of

FIGURE 6.1 Burned (a) and unburned (b) sugarcane areas with similar topographies, soils, and climate conditions.

sugarcane, influencing cane yields, weed control, fertilizer management, soil erosion, soil water infiltration rates, soil organic matter dynamics, and other factors. A brief discussion of the impacts of mulching in the unburned system is presented, focusing on soil organic matter, one of the key indicators of soil quality.

TOTAL SOIL CARBON

Several studies have indicated a trend for carbon sequestration under green cane management. The increase in stocks, though, is conditioned by factors such as climate, soil texture, nitrogen fertilizer management, and time since the adoption of the unburned

FIGURE 6.2 Trash accumulated on soil surface under unburned sugarcane in Pradópolis, southeastern Brazil.

harvest. Wood (1991), in trash management experiments in Australia, measured soil carbon content 20% higher in the 0 to 10 cm depth in areas without burned as compared to burned areas two years after the beginning of green cane management.

In a long-term study (55 years) comparing burned and unburned sugarcane in southeastern Brazil, Canellas et al. (2003) report C concentrations of 22.34 g kg^{-1} in the unburned cane, and 13.13 g kg^{-1} in the burned cane at the 0 to 20 cm soil depth. Razafimbelo et al. (2006) describe a 15% increase in soil carbon stocks in the 0 to 10 cm layer after six years of green cane management compared to management with burning. Vallis et al. (1996) describe a steady increase in carbon stocks in unburned plots, and no change in carbon stocks in a four-year period in adjacent burned areas. In southeastern Brazil, Feller (2001) reported that an average of 0.32 t C ha^{-1} yr^{-1} was accumulated in 12 years in the first 20 cm depth of an oxisol due to harvesting without burning. Luca (2002) reported annual soil carbon increases ranging from 1.2 to 1.9 t C ha^{-1} yr^{-1} for the 0 to 40 cm layer during the first four years of no burning.

Galdos et al. (2009a) measured carbon stocks changes in an oxisol under burned and unburned sugarcane and under an adjacent native vegetation area in Pradópolis, southeastern Brazil. Table 6.1 shows soil texture, bulk density, and pH data for the 0 to 10 and 10 to 20 cm layers. The two sugarcane fields with contrasting residue management had similar soil types, topography, and management. Both fields were planted eight years earlier and had not been plowed since then. Prior to planting, both areas were cropped with sugarcane for approximately 50 years, with preharvest burning. The soil total carbon concentration was determined by dry combustion. Total carbon stock was calculated from soil bulk density and total carbon concentration for each depth. To compare the same mass of soil, the stocks were corrected using the bulk density of the native forest as a reference. Figure 6.3 shows the corrected carbon stocks. The native forest area presented higher soil carbon stocks down to 20 cm, followed by the unburned sugarcane area and the burned area. The difference between burned and unburned soil carbon stocks was significant.

TABLE 6.1

Clay, Silt, and Sand Content, Bulk Density, and pH by Depth for Native Forest and Sugarcane Areas

Situation	Depth (cm)	Clay (g kg^{-1})	Silt (g kg^{-1})	Sand (g kg^{-1})	Bulk Density (g cm^3)	pH (H$_2$O)
Native forest	0 to 10	748	169	83	0.83 (0.02)	5.61 (0.10)
Native forest	10 to 20	792	143	65	0.93 (0.02)	5.48 (0.09)
Burned sugarcane	0 to 10	662	177	161	1.04 (0.02)	5.85 (0.09)
Burned sugarcane	10 to 20	681	176	143	1.16 (0.04)	5.98 (0.06)
Unburned sugarcane	0 to 10	738	174	89	1.05 (0.06)	5.39 (0.06)
Unburned sugarcane	10 to 20	750	159	91	1.06 (0.03)	4.92 (0.14)

Note: Standard errors indicated in parentheses.

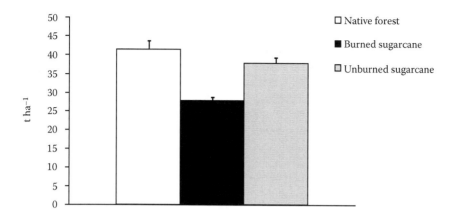

FIGURE 6.3 Soil carbon stocks (0 to 20 cm) in native forest and sugarcane areas eight years after replanting, corrected for density. Vertical lines on bars represent standard errors.

Time since adoption of the green cane management system exerts an impact on potential increase in soil carbon storage. In experiments in Australia and Brazil measuring carbon concentration under burned and unburned sugarcane, no significant difference between treatments was noted after 12 months (Blair et al., 1998). Robertson (2003) describes higher 0 to 10 cm soil C concentration in unburned areas in Australia after four to six years, but not in areas recently converted to green cane management (one to two years).

Another factor that affects soil carbon sequestration potential is initial carbon stock. In a long-term experiment in South Africa, the carbon concentration in the 0 to 10 cm soil layer was higher in unburned than in burned plots after 60 years, but no significant difference was found in the 10 to 20 and 20 to 30 cm layers (Graham et al., 2001). The carbon stock in the first 20 cm was high, 70 t ha^{-1}. This soil may have approached the limit of its capacity to store organic matter and reached equilibrium with respect to management (Six et al., 2002).

Another factor to consider is soil texture. Galdos et al. (2009a) found no significant differences in soil carbon stocks in the 0 to 10 cm layer in sugarcane fields four years after replanting with and without burning in Pradópolis, southeastern Brazil. The sugarcane areas were located in the same soil type on the same farm and showed marked differences in texture in the 0 to 10 cm layer: 57 g kg^{-1} sand in the burned field, and 351 g kg^{-1} sand in the unburned field. The strong correlation between soil carbon stocks and texture, mainly clay content (Silver et al., 2000) and clay-and-silt content (Hao and Kravchenko, 2007) is well-established.

Processed-based models are useful for assessing the potential for soil carbon sequestration in green cane management based on the limited number of long-term trash management experiments. Vallis et al. (1996), using the CENTURY model, estimated that maintaining trash on the soil surface could increase soil carbon stocks nearly 40% in ~70 years in Australia. The model was able to reflect the main trends in measured values, such as little change in soil carbon stocks in burned plots and steady increases in unburned plots. Galdos et al. (2009b) also detected long-term simulation trends of increased soil carbon sequestration under green cane management. The simulations were performed with data from experiments lasting one to 60 years in Goiana and Timbaúba, northeastern Brazil; Pradópolis, southeastern Brazil; and Mount Edgecombe, Kwazulu-Natal, South Africa (Figure 6.4). Prior to the long-term simulations, the model was validated against measured soil carbon data, with a high correlation ($R^2 = 0.89$). The long-term simulations confirmed the observed trend for higher carbon stocks in fields without preharvest burning of sugarcane. The simulation for the Mount Edgecombe site showed a decline in stocks in both trash management systems, although the decline was less pronounced in the unburned treatment, probably due to high initial carbon stocks in the 0 to 20 cm layer, around 70 t C ha^{-1}. Thorburn et al. (2002) found similar results for the same site in South Africa, using the soil carbon submodel of the APSIM Sugarcane model. They reported soil carbon decreases of 15% and 30% for the unburned and burned systems, respectively, in a simulation period of 60 years.

SOIL CARBON FRACTIONS

Microbial biomass carbon (C_{MB}), particulate organic matter carbon (C_{POM})—a recently added organic material with particles larger than 53 μm (Cambardella and Elliot, 1992)—and labile (permanganate-oxidizable) carbon have been shown to be more sensitive to sugarcane trash management than total carbon. Blair et al. (1998) found significant increases in the labile carbon fraction after green trash treatment compared to burning trash in the surface soils of two trash management trials in Australia. Similarly, Bell et al. (2001) showed that concentrations of labile carbon were more responsive to residue retention than total carbon. Increases in soil microbial biomass carbon content with green cane management were reported in several experiments (Wood, 1991; Sutton et al., 1996). Graham et al. (2001) described a significant increase in C_{MB} at the 0 to 30 cm depth in the unburned treatment in a long-term experiment in South Africa and noted no difference in soil total carbon content to this depth. Robertson (2003) describes an experiment indicating a significant increase in C_{MB} at the 0 to 10 cm depth in areas with four to six years of unburned harvest, but not in areas with one to two years of this management.

Galdos et al. (2009a) describe an experiment in southeastern Brazil comparing burned and unburned sugarcane areas with a reference native forest. Table 6.2 shows results for total carbon concentration, microbial biomass carbon, and particulate organic matter carbon. The C_{MB} content in the surface layer (0 to 10 cm) in the sugarcane areas was lower than in the native forest area ~50 years after conversion. In the 10 to 20 cm layer, the C_{MB} content of the sugarcane areas was lower, but the difference was smaller, with eight-year values in the unburned area of 265 mg kg^{-1}, close to the native forest carbon value of 268 mg kg^{-1}. The C_{MB} in the surface layer in the unburned area was 2.5 times higher than in the burned sugarcane area. In the 10 to 20 cm layer, the C_{MB} was 1.5 times higher in the unburned plot, compared to the burned system.

The soil particulate organic matter carbon (C_{POM}) in native forests was higher than in the sugarcane areas in both the 0 to 10 and 10 to 20 cm soil layers after a long-term

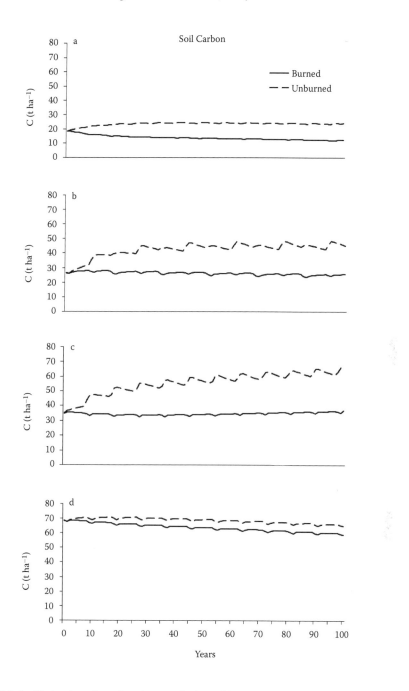

FIGURE 6.4 Projections for soil carbon stocks in a 100-year period comparing burned and unburned treatments at sites in Goiana (a), Timbaúba (b), and Pradópolis (c) in Brazil and Mount Edgecombe (d) in South Africa using CENTURY model.

TABLE 6.2

Total, Particulate Organic Matter, and Microbial Biomass Carbon in Native Forest, Burned, and Unburned Sugarcane Areas (0 to 10 and 10 to 20 cm Depths) Eight Years after Replanting

Situation	Total Carbon (g kg⁻¹)	Particulate Organic Matter Carbon (g kg⁻¹)	Microbial Biomass Carbon (g cm³)
		0 to 10 cm	
Native forest	30.12 (2.02) a	19.08 (1.25) a	618
Burned sugarcane	16.57 (0.45) c	2.44 (0.09) c	166
Unburned sugarcane	23.30 (0.91) b	9.27 (1.64) b	412
		10 to 20 cm	
Native forest	17.83 (1.04) a	7.13 (1.17) a	268
Burned sugarcane	15.40 (0.52) a	1.97 (0.10) b	180
Unburned sugarcane	18.70 (0.95) a	7.28 (1.17) a	265

Notes: Standard errors indicated in parentheses. Letters in columns indicate 5% significance difference for each depth.

sugarcane cropping history (Table 6.2). Silva et al. (2007), in a study measuring organic matter fractions in an oxisol cultivated with sugarcane, described a sharp decrease in C_{POM} right after conversion, followed by a slow increase. Nevertheless, after 25 years of cultivation, the C_{POM} was still lower than in native forest soil. The higher C_{POM} values were in the surface layer, as expected, since the organic matter addition was via plant reside placed on the surface without incorporation. The unburned sugarcane plot revealed higher C_{POM} values than the burned plot, with 2.44 g kg⁻¹ (burned) and 9.27 g kg⁻¹ (unburned), in the surface layer.

Table 6.3 presents the proportions of microbial biomass carbon in total carbon (C_{MB}/TC), and of particulate organic matter carbon in total carbon (C_{POM}/TC) for the native forest and sugarcane areas at 0 to 10 and 10 to 20 cm depths. In the 0 to 10 cm depth, the microbial biomass carbon represents ~2% of the total carbon in the native forest, with lower values for both sugarcane sites. This can be explained by the fact that C_{MB}, as a more labile pool, is lost first in a conversion from a native condition to a cultivated system. Dominy et al. (2002) describe experiment results in Australia in agreement with this work, with a decrease in C_{MB} in sugarcane areas cultivated for 20 to 30 years after conversion from forest. The microbial biomass carbon in the burned areas represented 1.0% (0 to 10 cm depth) and 1.2 % (10 to 20 cm depth) of the total carbon. The percentages were generally higher in unburned areas, 1.8 % (0 to 10 cm) and 1.4 % (10 to 20 cm) of the total carbon, confirming the effect of the maintenance of a litter layer on increases in the proportion of microbial biomass carbon fraction. Similar to the proportions found in microbial biomass carbon, the native forest had higher values of C_{POM}/TC than the sugarcane areas, especially in the surface layer. The unburned plots presented a higher percentage of soil particulate organic matter

TABLE 6.3

Proportions of Particulate Organic Matter and Microbial Biomass Carbon to Total Carbon in Native Forest Area, Burned, and Unburned Sugarcane Areas (0 to 10 and 10 to 20 cm Depths) Eight Years after Replanting

Situation	Particulate Organic Matter Carbon/Total Carbon (%)	Microbial Biomass Carbon/Total Carbon (%)
	0 to 10 cm	
Native forest	63.3	2.1
Burned sugarcane	14.7	1.0
Unburned sugarcane	39.8	1.8
	10 to 20 cm	
Native forest	40.0	1.5
Burned sugarcane	12.8	1.2
Unburned sugarcane	38.9	1.4

carbon in total carbon than burned areas. The proportions of C_{POM} in the burned area were 14.7% (0 to 10 cm) and 12.8% (10 to 20 cm); averages of the unburned areas were 39.8% (0 to 10 cm) and 38.9% (10 to 20 cm).

SOIL FAUNA

Other soil quality indicators are influenced by the maintenance of the litter layer in unburned sugarcane management. For example, the quantity and diversity of soil macrofauna are influenced as described in an experiment in southeastern Brazil by Cerri et al. (2004). Figure 6.5 presents the results of a comparison of macrofauna in a native forest, an area cropped with sugarcane for more than 50 years with preharvest burning, and an adjacent area with a long-term history of preharvest burning and harvested without burning for four years. The results show that more than 50 years under burned sugarcane caused a significant reduction in the number of individuals per hectare and in diversity.

In the burned sugarcane system, approximately 75% of the individuals were coleoptera larvae (sugarcane parasites). After four years of unburned harvesting, the number of individuals rose to levels above those found in native forests. The individuals that benefited most from the change from burning to maintaining a litter layer were ants and earthworms. Wood (1991) reported a threefold to fourfold increase in earthworm population after long-term trash maintenance (compared to burned sites) in a field experiment in Australia. Some of the ecosystem services provided by soil invertebrates are the comminution and incorporation of litter into soil, building and maintenance of structural porosity, aggregation through burrowing, casting and nesting activities, and control of microbial communities and activities (Lavelle et al., 2006).

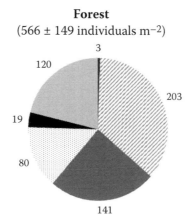

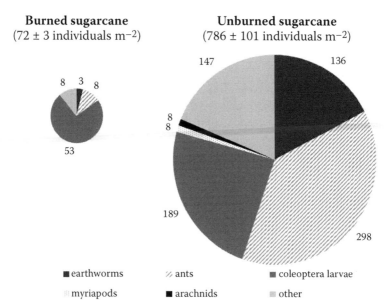

FIGURE 6.5 Number of microfauna individuals in 0 to 20 cm soil layer under forest, sugarcane burned before harvest for more than 50 years, and after four years of unburned harvest.

Replanting

Cultivation

Sugarcane is a perennial grass that needs replanting every six years on average. In preparation for crop renewal, the stool of the last harvested crop is killed and the soil is prepared for a new crop. The two primary systems for planting a new crop are conventional and minimum cultivation. Under conventional cultivation, the whole replant area is disturbed by tillage, disking, and sometimes subsoiling. Soil disturbance during replanting period when the soil has little vegetative cover can cause significant losses of topsoil (and thus organic matter), especially when coupled with erosive precipitation. Sparovek and Schnug (2001) estimated soil erosion rates of 31 t ha^{-1} yr^{-1} in sugarcane fields with preharvest burning and intensive tillage at replanting in a watershed in Piracicaba, Brazil.

Additionally, some of the carbon stored under the green cane management may be lost during replanting when the soil is tilled and disked, and sometimes subsoiled. A positive correlation between soil tillage and increased soil carbon mineralization was observed in several studies and attributed mainly to the destruction of soil aggregates and the subsequent exposure of organic material to decomposing biota (Six et al., 2002). Vallis et al. (1996), in an experiment with unburned sugarcane in Australia, reported a sharp reduction in carbon stocks after soil disturbance in the replanting period. Resende et al. (2006) in an experiment in northeastern Brazil, found no significant differences in soil carbon in the top 30 cm between burned and unburned cane fields after 16 years. They attribute the lack of differences to the heavy tillage implemented after eight years, when the first crop was plowed under and a new crop planted. Bell et al. (2001) reported significant increases in soil carbon in response to residue retention in the top 2.5 cm layer during late ratoon crops but no differences between burned and residue retained systems in plant crops, after land preparation.

COVER CROPS AND GREEN MANURES

Most sugarcane areas are seldom out of crop more than three to six months while the last ratoon is removed and land prepared for the next planting. Legumes have been cultivated in this interval, decreasing pest incidence by breaking the monoculture, providing protection against water erosion, increasing soil organic matter content with crop residue input, and providing biologically fixed nitrogen for the next sugarcane crop. A large variety of legumes have been used in the fallow period between sugarcane crops. Examples are soybean (*Glycine max*), peanut (*Arachis hypogaea*), cowpea (*Vigna unguiculata*), Dolichos lablab (*Lablab purpureus*), sun hemp (*Crotalaria juncea* L.), prickly sesban (*Sesbania aculeata*), and indigo (*Indigofera tinctoria* L.). Some fallow legumes such as peanuts and soybeans provide further benefit by serving as cash crops and providing extra income. Some recently implemented ethanol plants have been adapted for biodiesel production from soybeans grown in the fallow period, diversifying biofuel production and optimizing land use.

ORGANIC AMENDMENTS

In certain regions, it is common to apply residues from the processing of sugarcane as a method of cycling organic matter, nutrients, and water. Vinasse, also known as stillage, is a nutrient-rich liquid residue from distillation of ethanol. Between 13 and 15 L of vinasse are produced per liter of ethanol, and application rates can vary between 30 and 600 m^3 ha^{-1} yr^{-1}. According to Silva and Orlando Filho (1981), the average composition of vinasse in Brazilian ethanol plants was 8 to 50 g C L^{-1}, 0.1 to 1.5 g N L^{-1}, 0.02 to 0.77 g P L^{-1}, and 0.6 to 13 g K L^{-1} along with varying quantities of Ca, Mg, S, Fe, Cu, Zn, and Mn. Some of the benefits of applying vinasse on agricultural soils are increased nutrient availability, CEC and pH, improvements in soil water retention and structure, and enhanced soil microbial activity. However, the long-term application of vinasse may cause soil quality problems such as nitrate leaching and fertility unbalances, especially under high application rates.

Filter cake, also called filter mud or press mud, is a high fiber residue obtained by clarification of sugarcane juice. It can be a significant source of major plant nutrients like N, P, K, and other nutrients originating from the sugarcane plants and the chemicals used in cane juice clarification (Elsayed et al., 2008). Filter cake is usually applied at planting, in furrows or to the whole area, and can be mixed with other organic amendments. The application of filter cake leads to increases in soil C and N as well as to increased yields, as reported by studies in Sudan (Elsayed et al., 2008), India (Kaur et al., 2005), and Brazil (Rodella, 1990).

FUTURE DIRECTIONS

Ethanol can be produced from cellulose and hemicelluloses present in feedstocks, including fast-growing hays like switchgrass, short-rotation woody crops like poplar, and also from crop residues (Reijnders, 2008). While not cost-competitive today, advances in technology lead us to believe that in the next few years ethanol made from these feedstocks (second generation ethanol) will become commercially feasible. Considering the scarcity of arable land available to meet increasing energy demands, it is expected that crop residues will be totally or partially removed from fields for ethanol production. Blanco-Canqui and Lal (2007) noted that crop residues (corn stover, in their study) are indispensable for effective soil and water conservation. Excessive removal of stover for biofuel production may adversely affect soil organic carbon, nutrient cycling, soil tilth, water reserves, biotic activity, and crop yields (Lal et al., 2004). It may reduce soil water storage, alter soil temperature regimes, reduce structural stability, increase compaction, decrease water infiltration, and increase soil erosion and non-point source pollution.

Conversely, for sugarcane, partial removal of litter may even be beneficial, since in some soil and climate conditions, excessive mulch may hinder planting, fertilizer application and irrigation, increase disease and pest occurrence (e.g., *Mahanarva fimbriolata*), and delay ratoon emergence, resulting in lower crop yields. A compromise would involve leaving part of the residues on the field while processing the other part into bioethanol. This would still be beneficial to soils and crops, and allow production of renewable energy.

While the negative impacts of complete litter removal are foreseeable (Cerri et al., in preparation), the impact of partial removal on soil organic carbon, physical quality, and crop productivity has not been fully resolved. We know that crop residues applied to soil are important for soil organic carbon—an important determinant of soil fertility—and within limits, crop productivity is positively related to soil organic matter content. More data about maximum permissible rates of sugarcane litter removal under different soil and climate conditions must be obtained. Furthermore, it is essential to develop decision support systems for judicious management of crop residue for essential but competing uses since reduction of carbon levels in soils can contribute to increased levels of greenhouse gases in the atmosphere.

Considering the increased demand for alternative energy sources and pressing issues such as global warming and depletion of natural resources, the world needs sustainable biofuel production. Several current initiatives focus on the establishment of specific criteria for environmentally sound biofuel production including government standards and regulations. However, most criteria for the agricultural phase of biofuel production focus primarily on greenhouse gas emissions and energy use. Based on all the issues discussed in this chapter, soil quality should definitely be included in sustainability assessments for bioethanol.

REFERENCES

Bell, M.J. et al. 2001. Effect of compaction and trash blanketing on rainfall infiltration in sugarcane soils. *Proc. Austr. Soc. Sugar Cane Technol.* 23: 161–167.

Blair, G.J. et al. 1998. Soil carbon changes resulting from sugarcane trash management at two locations in Queensland, Australia and in Northeast Brazil. *Austr. J. Soil Res.* 36: 873–881.

Blanco-Canqui, H. and Lal, R. 2007. Soil and crop response to harvesting corn residues for biofuel production. *Geoderma* 141: 355–362.

Brandes, E.W. 1956. Origin, dispersal and use in breeding of the Melanesian garden sugarcanes and their derivatives, *Saccharum, Piracicaba officinarum* L. *Proc. Congr. Int. Sugarcane Technol.* 9: 709–750.

Canellas, L.P. et al. 2003. Propriedades químicas de um cambissolo cultivado com cana-de-açúcar, com preservação do palhiço e adição de vinhaça por longo tempo. Rev. Bras. Ciê. Solo 27: 935–944.

Cerri, C.C. et al. 2004. Canne à sucre et sequestration du carbone. Academie d'Agriculture de France.

Coelho, C.H. et al. 2008. Dissolved organic carbon in rainwater from areas heavily impacted by sugar cane burning. *Atm. Environ.* 42: 7115–7121.

Crutzen, P.J. and Andreae, M.O. 1990. Biomass burning in the tropics: Impact on atmospheric chemistry and biogeochemical cycles. *Science* 250: 1669–1678.

Dominy, C.S., Haynes, R.J., and van Antwerpen, R. 2002. Loss of soil organic matter and related soil properties under long-term sugarcane production on two contrasting soils. *Biol. Fertil. Soils* 36: 350–356.

Elsayed, M.T. et al. 2008. Impact of filter mud applications on the germination of sugarcane and small-seeded plants and on soil and sugarcane nitrogen contents. *Biores. Technol.* 99: 4164–4168.

FAOSTAT. 2009. FAO Statistical databases. http://faostat.fao.org/site/340/default.aspx>.

Feller, C. 2001. Efeitos da colheita sem queima da cana-de-açúcar sobre a dinâmica do carbono e propriedades do solo. Final Report Fapesp 98/12648-3, Piracicaba, Brazil.

Galdos, M.V., Cerri, C.C. and Cerri, C.E.P. 2009. Soil carbon stocks under burned and unburned sugarcane in Brazil. *Geoderma*, submitted.

Galdos, M.V. et al. 2009. Simulation of soil carbon dynamics under sugarcane with the CENTURY model. *Soil Sci. Am. J.* 73: 1–10.

Goldemberg, J. 2007. Ethanol for a sustainable energy future. *Science* 315: 808–810.

Graham, M.H. et al. 2001. Long-term effects of green cane harvesting versus burning on the size and diversity of the soil microbial community. *Proc. S. Afr. Sugar Technol. Assn.* 75: 228–234.

Hao, X. and Kravchenko, A.N. 2007. Management practice effects on surface soil total carbon: Differences along a textural gradient. *Agron. J.* 99: 18–26.

Intergovernmental Panel on Climate Change. 1995. *Climate Change in 1994: Radiative Forcing of Climate Change*. Cambridge: Cambridge University Press.

Kaur, K., Kapoor, K.K., and Gupta, A.P. 2005. Impact of organic manures with and without mineral fertilizers on soil chemical and biological properties under tropical conditions. *J. Plant Nutr. Soil Sci.* 168: 117–122.

Lal, R. et al. 2004. Managing soil carbon. *Science* 304: 393.

Lavelle, P. et al. 2006. Soil invertebrates and ecosystem services. *Eur. J. Soil Biol.* 42: S3–S15.

Le Blond, J.S. et al. 2008. Production of potentially hazardous respirable silica airbone particulate from the burning of sugarcane. *Atm. Environ.* 42: 5558–5568.

Luca, E.F. 2002. Matéria orgânica e atributos do solo em sistemas de colheita com e sem queima da cana-de-açúcar. PhD thesis, Universidade de São Paulo.

Marris, E. 2006. Sugar cane and ethanol: Drink the best and drive the rest. *Nature* 444: 670–672.

Mitchell, R.D.J., Thorburn, P.J., and Larson, P. 2000. Quantifying the immediate loss of nutrients when sugarcane residues are burnt. *Proc. Austr. Soc. Sugar Cane Technol.* 22: 206–211.

Razafimbelo, T. et al. 2006. Effect of sugarcane residue management (mulching versus burning) on organic matter in a clayey oxisol from southern Brazil. *Agric. Ecosyst. Environ.* 115: 285–289.

Reijnders, L. 2008. Ethanol production from crop residues and soil organic carbon. *Res. Conserv. Recycl.* 52: 653–658.

Resende, A.S. et al. 2006. Long-term effects of pre-harvest burning and nitrogen and vinasse applications on yield of sugar cane and soil carbon and nitrogen stocks on a plantation in Pernambuco, N.E. Brazil. *Plant Soil* 281: 339–351.

Robertson, F. 2003. Sugarcane trash management: Consequences for soil carbon and nitrogen. Final report. Nutrient Cycling in Relation to Trash Management Project.

Rodella A.A., Da Silva, L.C.F., and Filho J.O. 1990. Effects of filter cake application on sugarcane yields. *Turrialba* 40: 323–326.

Scharlemann, J.P.W. and Laurance, W.F. 2008. How green are biofuels? *Science* 319: 43–44.

Silva, G.M. de A. and Orlando Filho, J. 1981. Caracterizacao da composicao quimica dos diferentes tipos de vinhaça no Brasil. *Bol. Tecn. Planalsucar* 3: 5–22.

Silva, A.J.N. et al. 2007. Impact of sugarcane cultivation on soil carbon fractions, consistency limits and aggregate stability of a yellow latosol in Northeast Brasil. *Soil Tillage Res.* 94: 420–424.

Silver, W.L. et al. 2000. Effects of soil texture on belowground carbon and nutrient storage in a lowland Amazonian Forest ecosystem. *Ecosystems* 3: 193–209.

Six, J. et al. 2002. Stabilization mechanisms of soil organic matter: Implications for C saturation of soils. *Plant Soil* 241: 155–176.

Sparovek, G. and Schnug, E. 2001. Temporal erosion-induced soil degradation and yield loss. *Soil Sci. Soc. Am. J.* 65: 1479–1486.

Sutton, M.R., Wood, A.W., and Saffigna, P.G. 1996. Long term effects of green cane trash retention on Herbert River soils. In *Sugarcane: Research toward Efficient and Sustainable Production*. Wilson, J.R. et al., Eds. Brisbane: CSIRO, pp. 178–180.

Thorburn, P.J. et al. 2002. The impact of trash management on soil carbon and nitrogen I: Modelling long-term experimental results in the South African sugar industry. *Proc. S. Afr. Sugar Technol. Assn.* 76: 260–268.

Vallis, I. et al. 1996. Simulation of the effects of trash and N fertilizer management on soil organic matter levels and yields of sugarcane. *Soil Tillage Res.* 38: 115–132.

Wood, A.W. 1991. Management of crop residues following green harvesting of sugarcane in north Queensland. *Soil Tillage Res.* 20: 69–85.

7 Economic Balance
Competition between Food Production and Biofuels Expansion

Jozsef Popp

CONTENTS

GLOBALIZATION AT RISK

The sustained economic growth worldwide during the past two decades has shown the benefits of globalization. Although it must be admitted that much more could have been achieved if more progress on the Doha Development Agenda on trade had been made. However, the current lower growth worldwide, rising unemployment, and deflating asset values may trigger nationalism, excessive self-interest, and protectionism. We need more responsibility for fairer world trade to avoid globalization that enriches a few beyond belief and excludes most of the population. Trade responsibility also means accepting special and differential treatment for developing countries and giving them temporary trade protection so they may protect themselves from food import surges.

We face a future of food scarcity, with high, albeit very volatile, prices both for inputs and outputs. Food scarcity is aggravated by managed trade, and lack of financial stability, and consequent environmental degradation. Recent events have proven that markets can fail as deregulation has backfired. Liberalized trade and related financing depend on stable markets. A new financial architecture is urgent. We also

151

need greater responsibility in budgetary and financial affairs to avoid passing the burden of adjustment to future generations.

More responsibility is needed related to food trade, supporting a coordinated regulatory framework, and ethical public and private behavior in fighting environmental degradation. We need greater responsibility in cutting greenhouse gas (GHG) emissions, showing greater respect for the environment, and strengthening and widening the Kyoto process. Southern hemisphere countries must introduce land reforms, allowing the poor to access the land, and adopt more appropriate food pricing policies. To produce sufficient food at prices affordable for everyone, we may also have to change our food habits. The reform process of the Common Agriculture Policy (CAP) must be continuously adapted to changing realities.

Can we stop excessive ecological and financial borrowing from affecting future generations? The current crisis has proven that governments can act decisively and even effectively if extraordinary circumstances so dictate; we are not short of medicine (i.e., solutions, capacity, funds), but we need political will to apply it— and usually have it only when a crisis has manifested itself. The "costs of earlier inaction" will rise. So how can we prepare ourselves? Only true political leadership and effective institutions will change the current situation.

GLOBAL RESPONSIBILITIES FOR FOOD, ENERGY, AND ENVIRONMENTAL SECURITY

The current global economic crisis may well become the longest in three generations. Investment and trade protectionism may be temporarily on the increase. If trust in finance and economy does not return rapidly, economic reform, socio-economic growth and political stability will suffer. While some confidence in the financial system will return in due course, a new financial architecture is required to strengthen the global economy and increase economic and financial fairness. In this connection, it is critical that the needs for global food and environmental security are taken into account, and that ways are found to finance the services that farmers provide to society.

World population growth is the biggest trend-making factor: 70 to 80 million more people a year, close to 8 billion by 2025. Population growth creates a rapidly growing demand for crop products including feed arising from increasing meat consumption. Other major global trends are globalization and urbanization. Moving production to the most competitive regions causes the food trade to become more liberalized and also more concentrated. Growing energy demand and climate change will also influence food production; agriculture will contribute to emissions into the environment and also suffer or benefit from changing climates, depending on climatic zones. Additional challenges are increasing market volatility resulting from yield and end stock fluctuations and consumer sensitivity to food quality, safety, and price. Uncertainty surrounds the timing and application of innovations in biotechnology, nanotechnology, precision farming, carbon sequestration, and information technology. Finally we face the question of who will pay for agricultural public services provided by land managers that the market does not pay for, such as rural landscape maintenance, environmental protection biodiversity, and animal welfare. These challenges are aggravated by global irresponsibility related to food security, water and environmental sustainability—and energy security.

FOOD SECURITY

The recent food crunch may have eased somewhat. Wheat, corn, and rice prices tumbled 40% to 60% from all-time highs. In the European Union (EU), average producer prices increased by 2.7% in 2008, while agricultural input prices shot up by 11.6% (plus 60.6% for fertilizers and soil improvers). In 2008, the world's food import bill surged above $1 trillion—23% more than in 2007 and 64% more than in 2006. Developing countries spent in 2008 about one-third more of the world's food bill more than in 2007. Latin America, Africa, and Eastern Europe present good potential for new land cultivation, but new land is insufficient and inappropriate because of poor or polluted soils or difficult to use for food production because of doubtful property rights, poor financing, government mismanagement, and lack of transportation infrastructure. Moreover, cultivated land is diminishing fast because of urbanization and because of desert expansion. The addition of some 70 million people every year claims nearly 3 million hectares to provide housing, roads, and parking lots. The main reasons for diminishing food supplies are population growth and accelerated* urbanization, changes in lifestyles, falling water tables, and diversion of irrigated water to cities (The Earth Institute, 2005). All these factors generate losses in soil availability, quality, and crop use.

To meet world demand, the necessary production growth will to a large extent have to be met by a rise in the productivity of land already farmed today. However, this will be difficult to accomplish because global agricultural productivity growth has been in decline since the Green Revolution of the 1960s and 1970s. Global crop yield increases plummeted from 4% per annum in the 1960s to 1980s to 2% in the 1990s and barely 1% in forecasts for 2000 to 2030. Despite substantial expected yield increases in India, United States, Russia, and the Ukraine, Europe's role as provider of food to the world is diminishing. The net crop trade position of the EU-27 can be expected to deteriorate. The EU's capacity to help fight world starvation will be reduced at a time when food production will decline steeply in countries that already face increasing food import needs. Nevertheless, Europe will become a more secure production location in comparison to other world regions and higher food prices will boost deforestation there.

All countries must improve their food security policies, in particular in Africa. Consumption patterns will change by increasing consumption of meat and dairy products. According to the Food and Agricultural Organization (FAO) of the United Nations, the race by some countries to secure farmland abroad risks creating a neo-colonial system (FAO, 2008). For example, Daewoo Logistics of South Korea, the fourth largest maize importing country, seeks to lease vast areas of productive agricultural land in Madagascar and elsewhere to satisfy domestic demand for maize and

---

* An estimated 40,000 ha of land are needed for basic living space for every 1 million people added.

palm oil. There are precedents starting with Japan's purchase of land in the American West to produce beef for Japanese consumers. Ethiopia, a poor country infamous for more frequent food crises, has just openly offered Middle Eastern countries leases of hundreds of thousands of hectares of its farmland to help them ensure food security in the Middle East. These plans show how important food security has already become. Political problems may eventually arise if a country hosting foreign investment in farming faces a serious food crisis at a time when rich foreigners export all the food produced for the exclusive benefit of richer and better fed people abroad.

The discussion of food crisis has faded into the background—overshadowed by the global macroeconomic crisis and the financial crisis. The sharp rise in prices of basic foodstuffs created extreme difficulties for a large part of the world's population. The food crisis affected more people more severely than the macroeconomic issue because the populations most affected by sharply rising food prices spend larger shares of their income on food. The global food crisis produced an extraordinary human impact, larger and more adverse than the global financial crisis. One indication of the severity is the remarkable amount of recent civil unrest and political instability in dozens of countries (Ethiopia, Egypt, Mexico, Thailand) because people were unable to afford basic nutrition.

Political responses were also extraordinary. Much of the world's system of trade in foodstuffs broke down temporarily as food exporting countries moved to limit or even ban exports in attempts to provide some protection to their domestic consumers. The severe economic slump worldwide represents an extraordinary world downturn—the worst downturn since the great depression. All these issues have diverted attention from the food crisis. The macro-economic crisis led many people to write off the food and more broadly the commodity price crisis of 2008 as a widespread belief that the event was a speculative bubble—too many people traded commodities, driving commodity prices to unsustainable levels—and that concerns about ultimate supplies of food were misplaced (Krugman, 2009).

International trade in commodities futures expanded enormously and food prices increased very sharply. Commodity prices rose very sharply and then fell precipitously. The assumption that the situation was a speculative bubble is incorrect. The rise and fall of commodity prices affected not only commodities with large futures but those without, for example, iron ore. Trading commodity futures only affects the price to the extent that speculation leads to withdrawal of real supplies and subsequent hoarding. However, that was not the case with agricultural commodities because food stocks were at record lows. During an economic slump, the real prices of commodities always fall and vice versa. The great depression showed a spectacular collapse of agricultural prices. The current fall in prices is a consequence of a global recession.

The end of crisis resource constraints along with ineffective polices are creating major problems for the world food supply. Despite sharp falls in food prices since their peak in early 2008, prices of basic foodstuffs are still higher than they were in 2000. Along with the continuing upward trend of food prices, volatility is a clear problem. People do not eat at long intervals; they eat every day. Should high prices from 2008 return, the problem will be very serious because people are very vulnerable to high prices. Whenever a country imposes an export ban, the global economy is affected even if its domestic consumers are protected.

The poor have no ways to diversify risk and no protection against high food prices. What can be done? One resolution is to invest in future food production both by planning and planting and through research and development (R&D). We tend to think of agriculture as a one-on-one economic exercise: market producers and consumers setting the market right. This is true but only to a point. Agricultural production and progress depend heavily on public goods and activities, especially R&D. Research and the agricultural infrastructure have attracted less emphasis in recent years largely because people thought the food issue was resolved. In reality, we have seriously underinvested and must "play catch up" (Krugman, 2009).

What about stability? The international community scrambled to provide financial aid to countries that suffered severely from high food prices. Aid helped, but finding the money was difficult. Looking back at history, much aid to people in Africa came from Saudi Arabia. Based on current oil prices, this may not happen again. We should have a reserve system in place for emergency support. The ad hoc response appears better than financial aid.

Many governments went too far from the policy of maintaining domestic food stocks as protection against crisis. Global markets function well when an individual country's harvest fails, but a systemic hike in food prices leads to global market breakdown. Food is different from other markets such as steel. Food is fundamental to maintain life, particularly in poor countries. Global food price rises cause global markets to break down exactly when we need them most.

Today we face severe slumps in the economy; this means a decline in relative prices of raw materials. With the end of recession, we again face a growing population, growing purchasing power, and growing consumption of foods that require very intensive production of cereals, for example, meat requires a lot more agricultural inputs than consumption of grain. Water and use of potential arable land are other concerns. When arable land is diverted to non-agricultural uses, it usually raises GDP, but it also reduces the incomes of those already at the bottom of the earning scale.

We saw a very serious outbreak of human suffering and political instability resulting from a very brief spike in the price of food. The spike did not continue for an extended time, and was overshadowed by a broad collapse of economic activity thanks to the financial crisis. Had the food price spike continued longer, it might have been much worse, and all indications are that the crisis of 2008 was a dress rehearsal for future crises, and no mechanisms to deal with such problems are in place.

ENERGY SECURITY

Energy prices have seen a decline (in constant dollars) over the past 200 years. The latest energy price hikes have not even brought us back to the price levels of some 30 years ago. The tragic reality is that political zeal led governments to keep energy prices as low as possible, thus frustrating most attempts to increase energy productivity. Energy price elasticity is very much a long-term affair, and return on infrastructure investments crucial to the creation of an energy-efficient society requires time.

Creating a long-term trajectory of energy prices that slowly, steadily, and predictably rise in parallel with energy productivity would give a clear signal to investors and infrastructure planners that energy efficiency and productivity are both necessary and profitable.

Much debate surrounds the potential contribution of agriculture to renewable energies. Unfortunately, existing technologies produce energies that may be renewable, but most are not green. Whether second generation biofuels may eliminate most of the pitfalls of the first generation is open to doubt, although they include saving food components of plants. Biofuel policy is a major aggravating factor even if it is now in the background because of the decline in oil prices that reduced the demand and the drops in food prices. The current economic crisis is now the focus of attention, but renewable energy will return as a problem when the macro-economic crisis ends.

Biofuels

Bioenergy represents approximately 10% of total world energy supply. Traditional unprocessed biomass accounts for most production, but commercial bioenergy is assuming greater importance. Liquid biofuels for transport garner the most attention and production has expanded rapidly. However, their quantitative role is only marginal: They constitute only 1% of total transport fuel consumption and 0.2% to 0.3% of total energy consumption worldwide.

The main liquid biofuels are ethanol and biodiesel. Both can be produced from a wide range of feedstocks. The most important producers are Brazil and the United States for ethanol and the EU for biodiesel. Current technologies for liquid biofuels rely on agricultural commodities as feedstocks. Ethanol is based on sugar or starchy crops; sugarcane in Brazil and maize in the U.S. produce the most significant volumes. Biodiesel is produced from a range of different oil crops.

Large-scale production of biofuels implies large land requirements for feedstock production. Liquid biofuels can therefore be expected to displace fossil fuels for transport to only a very limited extent. Although liquid biofuels supply only a small share of global energy needs, they still have the potential to significantly affect global agriculture and agricultural markets because of the volume of feedstocks and the land areas needed for their production.

The contributions of different biofuels to reducing fossil-fuel consumption vary widely when the fossil energy required for their production is taken into account. The fossil energy balance of a biofuel depends on factors such as feedstock characteristics, production location, agricultural practices, and the source of energy used for conversion. Different biofuels also perform very differently in terms of reducing GHG emissions. Second-generation biofuels currently under development will use ligno-cellulosic feedstocks such as wood, tall grasses, and forestry and crop residues. This would increase the quantitative potential for biofuel generation per hectare of land and may also improve the fossil energy and GHG balances of biofuels. However, it is not known when such technologies will achieve significant commercial-scale production.

Liquid biofuels such as bioethanol and biodiesel compete directly with petroleum-based petrol and diesel. Because energy markets are larger than agricultural markets, energy prices tend to drive the prices of biofuels and their agricultural feed-stocks. Biofuel feedstocks also compete with other agricultural crops for productive resources; therefore, energy prices will tend to affect prices of all agricultural commodities that rely on the same resource base. For the same reason, producing bio-fuels from non-food crops will not necessarily eliminate competition between food and fuel. For all technologies, the competitiveness of biofuels will depend on the relative prices of agricultural feedstocks and fossil fuels. The relationship will differ among crops, countries, locations, and technologies used in biofuel production.

With the important exception of ethanol produced from sugarcane in Brazil, which has the lowest production costs among the large-scale biofuel-producing countries, biofuels are not generally competitive with fossil fuels without subsidies. At current low crude oil prices, ethanol production even in Brazil is not competitive with gasoline. However, competitiveness can change based on changes in feedstock and energy prices and technology developments.

Biofuel development in developed countries has been promoted and supported by governments through a wide array of policy instruments; a growing number of developing countries are also beginning to introduce policies to promote biofuels. Common policy instruments include mandated blending of biofuels with petroleum-based fuels and subsidies. The exact contribution of expanding biofuel demand to these price increases is difficult to quantify. However, biofuel demand will continue to exercise upward pressure on agricultural prices as oil prices increase.

Modern bioenergy represents a new source of demand for farm products. At the same time, it generates increasing competition for natural resources, notably land and water, especially in the short run, although yield increases may mitigate such competition over time. Competition for land becomes an issue, especially when certain crops (maize, oil palm, and soybean) now cultivated for food and feed are redirected to the production of biofuels and also when food-oriented agricultural land is converted to biofuel production.

Biofuel policies present significant implications for international markets, trade and prices for biofuels, and agricultural commodities. Current trends in biofuel production, consumption and trade, and the global outlook are strongly influenced by existing policies, especially those implemented in the EU and United States that promote biofuel production and consumption while protecting domestic producers (Figure 7.1).

Trade policies vis-à-vis biofuels discriminate against developing country producers of biofuel feedstocks and impede the emergence of biofuel processing and exporting sectors in developing countries. Many current biofuel policies distort biofuel and agricultural markets and influence the location and development of global industries such that production may not occur in the most economically or environmentally suitable locations. International policy disciplines for biofuels are needed to prevent repeats of the global policy failure in the agriculture sector.

Currently, about 80% of the global production of liquid biofuels is in the form of ethanol. In 2008, global ethanol fuel production reached 65 billion liters and global biodiesel production amounted to 13 million tons. The two largest ethanol producers, Brazil and the United States, account for 85% of total production, with the remainder accounted for mostly by the EU (mainly France and Germany) China, and Canada

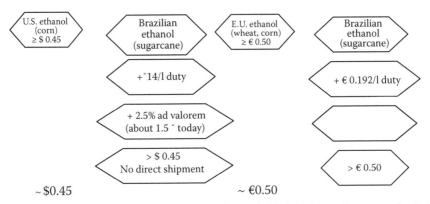

Rotterdam cif (T1): $0.43/l (€0.33/l) + €0,192/l duty = €0.51/l (ethanol price in the EU is largely determined by the exports from Brazil)
Rotterdam fob inc. duty: €0.51/l

FIGURE 7.1   Ethanol trade in European Union and United States. (*Source:* Licht, F.O., 2009, and authors' calculations.)

(Figure 7.2). Biodiesel production is concentrated in the EU (about 55% of the total), with a significantly smaller contribution coming from the United States. In Brazil, biodiesel production is a more recent phenomenon and production volume remains limited. Other significant biodiesel producers include Argentina and to a lesser extent India, Indonesia, and Malaysia (Figure 7.3). Brazil, the EU, and United States are expected to remain the largest producers of liquid biofuels, but production is also projected to expand in a number of developing countries.

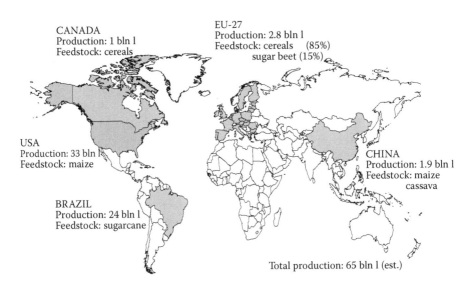

FIGURE 7.2   Word fuel ethanol production, 2008. (*Source:* Licht, F.O., 2009, and authors' calculations.)

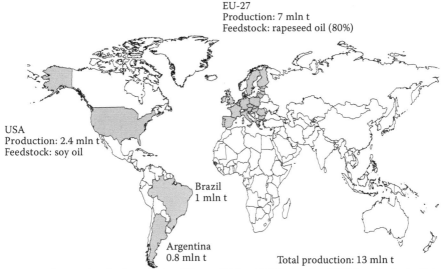

EU-27
Production: 7 mln t
Feedstock: rapeseed oil (80%)

USA
Production: 2.4 mln t
Feedstock: soy oil

Brazil
1 mln t

Argentina
0.8 mln t

Total production: 13 mln t

Note: Ethanol and crude oil parity prices (February 2009): at €0.50/l ethanol and $103/b crude oil price (but crude oil price was $44/b).

FIGURE 7.3   World biodiesel production, 2008. (*Source:* Licht, F.O., 2009, and authors' calculations.)

Feedstock Use

An increase in use of grains for fuel ethanol is expected by 2009, mainly due to a higher output in the United States and Europe. The increase could exceed 2008–2009 grain consumption by 6% (5.6% for previous season). Net grain use for fuel ethanol is one-third lower (4%), as ethanol yields dried distiller grains (DDGs) as by-products. The bulk of the worldwide use of grains in alcohol production involves corn in the United States and China. However, an increase in the offtake of wheat for fuel ethanol can also be observed in Canada and the EU. Due to the limited growth prospects for biodiesel, only a small rise in vegetable oil use for biodiesel manufacture can be foreseen and may even fall to 8.9% from 9.0% as non-fuel vegetable oil consumption inceases at a faster pace (Licht, 2009).

One of the key issues in the sustainability debate is the use of palm oil for biodiesel manufacture. Oil palm plantations are blamed for reducing ecologically precious areas in Southeast Asia and contributing significantly to increased GHG emissions. Nevertheless, palm oil may have become attractive for making biodiesel due to its recent low prices. However, at best, 6% of international palm oil demand (~2.6 million tonnes) will end up in the biodiesel sector in 2009. Domestic consumption in Malaysia and Indonesia remains low due to the lack of blending obligations. In 2009, the use of imported palm oil for biodiesel in the EU-27 may rise by 100,000 tonnes to 700,000, against a total biodiesel output of more than 7 million tonnes.

A decrease in global soybean oil usage for biodiesel manufacture in 2009 is projected, mainly based on a lower biodiesel output forecast for the United States arising from the end of the soy oil methylester (SME) B-99.9 trade. Growth prospects for Argentina are limited, and increases in biodiesel production in Brazil may not be large enough to offset reductions in the United States. Based on declines in vegetable oil values, the volume of used cooking oil and animal fat may fall slightly. In addition, a portion of biodiesel export from the United States to Europe was made from used cooking oil and animal fat. Nevertheless, these feedstocks may benefit significantly from the forthcoming sustainability decrees in the EU.

The share of fuel ethanol in the United States corn balance is growing. In 2008–2009, ~30% of the total corn demand will come from the fuel alcohol sector. Under the Energy Independence and Security Act, 2009 fuel ethanol use from corn is to reach 10.5 billion gallons and rise to 15 billion after 2015. However, the decline in gasoline prices changed the economics and may limit growth in corn use for fuel consumption. The expected increase in the EU's fuel alcohol output in 2009 will also boost feedstock consumption. The number of EU facilities capable of processing both wheat and sugar beets is growing. Due to firming prices, the use of molasses may fall in 2009 with the sugar co-product used mainly by fuel ethanol manufacturers in Hungary and Italy.

South America's ethanol sectors may see increases in cane and molasses processing. In Brazil, moderate growth is expected in fuel alcohol production. Increases in 2009 sugar output may limit growth in the usage of cane for ethanol. Thai ethanol producers mainly process molasses at present. The government plans to increase cane yields and reduce the cane processed into sugar for the export market. The use of tapioca is also seen rising as the government plans to boost production by attractive intervention prices. Tapioca plantings grow mainly at the expense of corn.

Total worldwide growth in use of vegetable oils for biodiesel is expected to shrink. The EU's biodiesel industry depends greatly on vegetable oil imports. EU-27 balances for rapeseed, rapeseed oil, soybeans, and soy oil—the major energy crops for biodiesel in Europe—show that a significant share of EU demand is imported from Third World countries. Rising biodiesel output increases vegetable oil demand while the feedstock potential from alternative non-food feedstocks such as animal fat and used vegetable oils is limited. Growth of domestic production through acreage expansion is also limited since demand for land to grow other crops such as grains is unlikely to decrease. The Ukraine may be a potential supplier for the EU vegetable oil market. The Ukraine already exports a high proportion of its rapeseed output. Production has grown, mainly on improved export prospects, since domestic demand is relatively small.

The EU is set to remain the largest biodiesel producer and thus the main consumer of vegetable oils for fuels, but growth rates are also declining because of lower fuel prices. Around 50% of EU-27 biodiesel output in 2009 will come from rapeseed oil, the equivalent of more than 9 million tonnes of the oilseed; 2008 rapeseed production was 19 million tonnes. Imports for 2009 may reach ~1.6 million tonnes. Soy oil demand for this use, roughly one-third of 2009 output, will be met totally by imports (as beans or oil). The share of palm oil in 2009 may increase on the back of lower values, but a significant share of the potential palm oil methyl ester (PME) offtake may well be met by imports of finished product. Due to its higher cold filter plugging point, PME is only usable in summer months or to a limited extent in winter months.

Lower 2009 U.S. biodiesel output is a major cause for the decrease in global soy oil usage to manufacture fuel. From the 2008–2009 U.S. production of the oil, 12% at most may be used for the fuel, down from 14% in the prior season. At the time of the recent rally in commodity prices, the U.S. sector showed a remarkable processing flexibility; up to 50% of monthly output came from non-soy feedstocks. Inedible tallow and grease along with rapeseed and palm oil accounted for most non-soy materials. However, soy oil is set to remain the main biodiesel feedstock in the U.S. in the absence of restrictions related to sustainability issues. The recent hike in values underscored the need for this sector to have flexible processing capacities.

Due to a lack of mandates, biodiesel is not that important for the Malaysian and Indonesian palm oil sectors. Total manufacturing capacity in these two countries is 2.8 million tonnes annually. Output in 2009 is expected to reach 950,000 tonnes, up from 550,000 in 2008 due to the cessation of SME B-99.9 business. That compares with seasonal palm oil exports of around 30 million tonnes (Licht, 2009).

The main challenge of the biofuels industry is how to cope with low fuel prices. The longer term outlook for fuel prices remains bullish. The question for the biodiesel sector will be how many companies are capable of surviving hard times? An adjustment in production capacity seems inevitable, and manufacturers belonging to conglomerates or integrated in the value chain usually have better chances.

Outlook for Fuel Ethanol Production

Fuel ethanol production in the United States reached a massive 33 billion liters (9 billion gallons) in 2008. In 2008, Brazil shipped 2.8 billion liters (740 million gallons) of ethanol either directly to the U.S. or through CBI* countries. Whether Brazilian alcohol can be mobilized for U.S. use depends primarily on price. Direct exports of anhydrous ethanol are out of the question now that the re-export loophole in customs regulation was closed by the latest farm bill. Keep in mind also that Brazilian millers have little incentive to produce more than minimum volumes of ethanol as long as sugar trades at the current premium to green fuel.

In 2009, CBI countries can ship 2.35 billion liters (622 million gallons) of ethanol duty-free to the U.S. In this case, the economics are more compelling. Whether the CBI quota will be sufficient to fill the gap on the American market remains to be seen. If not, ethanol values must rise considerably to attract additional volumes from domestic or overseas sources. In the EU, total bioethanol production in 2008 was 2.8 billion liters. Bioethanol imports increased by 400 million liters to almost 1.9 billion, of which 1.4 to 1.5 billion came from Brazil. Around 50% of the total was used in the fuel sector (Licht, 2009).

The year 2009 will be a defining one for the U.S. ethanol sector. A combination of relatively high corn prices and rock-bottom gasoline values threatens the industry. Higher grain costs put margins under pressure, and then the meltdown in the financial markets prompted gasoline prices to tumble. The industry is getting nervous, and the consensus that carried it through all previous crises is wearing thin. With a new administration now in place in Washington, the American renewable

---

* The Caribbean Basin Initiative (CBI) trade program is intended to facilitate the economic development of the Caribbean Basin economies.

fuels sector has new hope. President Obama has already made it clear that renewable energy will play a key role in his economy stimulus package. Biofuels gained little of substance from the American Recovery and Reinvestment Act (ARRA), but the boost may have come too late in 2009 to benefit the industry. The trend in gasoline prices is critical. Distillers cannot expect an improvement in their cost structures any time soon. The collapse of oil prices benefited American motorists much more than those in countries where taxes constitute higher proportions of retails prices. Lower values have made all types of alcohol uncompetitive in the U.S.

Brazil is expected to produce 26.5 million tonnes of sugar and 23.9 billion liters of ethanol in 2009. In the 2007–2008 season, it produced 26.2 million tonnes of sugar and 20.3 billion liters of ethanol, according to the Sugar Cane Industry Association (Unica). In contrast to expectations, mills to date have not been able to export more than 15% of the fast growing ethanol production. In 2008, almost two-thirds of Brazil's ethanol exports went to the U.S., some via CBI countries. These countries can re-export up to 2.35 billion liters of dehydrated alcohol to the U.S. in 2009 free of the high duty imposed on ethanol imported directly from Brazil. Before oil values collapsed last year, alcohol imported directly from Brazil was competitive with gasoline, even after payment of high duties. In addition, some oil firms took advantage of a loophole that allowed tax-free import of ethanol on a "draw-back" scheme if an identical amount of some other fuel was exported—a practice halted at the end of September 2008 (Licht, 2009).

This proportion of export will fall to little more than 10% of output in 2009–2010, when fewer than 3 billion liters will be sold overseas, compared with the 4.6 billion liters in 2008–2009. Exports will probably remain around that level until oil prices start to rise again. Contrary to hopes, no other market for ethanol similar to the size of the U.S. market has emerged, nor does any seem likely to appear for some time, and the development of large-scale trade with Japan remains a pipedream. On the other hand, the EU's determination that 10% of motor fuels consumed within the Community must be renewable after 2020 should also favor producers. Brazil has a good chance to supply a large chunk of the 14 billion liter market that may well develop as a result of the requirement. Although developments in the United States and EU appear beneficial for alcohol in the long term, the sector in Brazil will face extremely difficult times until that happens.

Many leading firms reported substantial losses for 2008. While prospects for sugar are fairly bright, those for ethanol remain gloomy. The economy is expected to grow by 2% at most in 2009, compared with 5.5% last year. With unemployment rising fast and causing consumer caution, motorists will use much less alcohol than they have in recent years. The current crisis is one of the most serious ever experienced by Brazil's sugar and alcohol industry. Access to credit for operations and exports has been cut at a time when the sector is halfway through a costly expansion program and low values for both commodities have weakened many companies.

Brazil's $50 billion expansion plan and its attraction for investors were based on the assumption that falling supplies of crude oil along with likely price rises would lead to steady growth of world demand for ethanol. However, this has not happened and the United States has been Brazil's only significant overseas market to date. Anticipating increasing offtakes, many Brazilian companies borrowed heavily in U.S. dollars to pay for expansion programs. Borrowing abroad made good financial sense while the Brazilian real was gaining ground against the U.S. dollar. However, now that the real has lost a third of its value, servicing U.S. dollar denominated debts has become a great burden. Although it proved possible to dispose domestically of the huge amounts of extra alcohol produced, intense competition kept values below the cost of production for most of 2008. With little prospect of improvement of the international ethanol scene, firms are hoping that increasing sugar consumption will help save the sector.

Based on low sugar values and strong demand for ethanol, the proportion of cane distilled into alcohol exceeded 60% in 2008. This trend will be reversed in 2009–2010, partly because much less extra alcohol will be needed this year and partly because a world deficit of 3 to 4 million tonnes of sugar is expected in 2009, taking account of the difference between supply and demand. The international sugar price has increased to reflect this. The decline of the Brazilian real by more than 30% against most foreign currencies since the middle of 2008 means that producers will receive more in the currency in which most of their costs are incurred.

Because it costs less than $40 to make the ethanol equivalent of a barrel of oil, it was anticipated that green fuel would become steadily more competitive and popular, and consequently the demand would continue to grow. This scenario still holds true, which explains why many investors postponed their plans but have not abandoned them. The current difficult phase may last for some time. However, once the economies of enough countries will grow fast enough to transform the present surplus of oil into a shortage again, the price of oil will quickly rise above $100 per barrel.

EU's continued commitment to 10% mandate for 2020 is welcomed. The package will require the EU to derive 20% of its energy from renewables by 2020, including 10% of its transportation energy, mostly from biofuels. Starting in 2010, biofuels must achieve GHG savings of 35% relative to fossil fuels; this must rise to 50% by 2017. Biofuel plants beginning operation in 2017 must achieve savings of 60%. Biofuel consumption in Eastern Europe is set to rise on the back of increasing biofuel mandates. A significant share of this demand will be met by domestic production. To a growing extent, markets in the new member states (EU-12) will have to compete with EU-15 and non-community imports. The EU does not lack biofuel production capacity. This limits export prospects for ethanol projects with similar cost structures outside the EU. Competitiveness of ethanol production depends on the relative prices of feedstock and fossil fuel (Figure 7.4).

In Asia, biofuels in general and ethanol in particular have been introduced to alleviate the chronic energy shortage that dogs many of the region's economies. With crude oil prices falling below $70 a barrel, the need to develop domestic sources of energy has lost some of its urgency. While lower commodity values seen recently have reduced production costs for ethanol, this fall has not been sufficiently large to compensate for the sharp decline in crude oil prices.

Thailand has been promoting biofuels with a comprehensive package of policy measures since 2003. Its distilleries are presently working below capacity due to limited overseas opportunities and disappointing domestic gasohol demand. The current excess of supplies will prevent short-term expansion of alcohol capacities. Traditionally, China has used grains to manufacture fuel ethanol. Currently, four of the five plants in the country use cereals; only one uses tapioca starch. The use of this

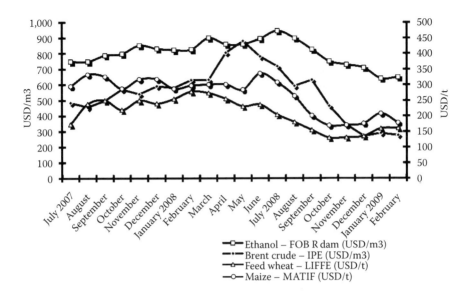

FIGURE 7.4   Prices of ethanol, crude oil, feed wheat, and maize in European Union, July 2007 through February 2009. (*Source:* HGCA, Kingsam and author's calculations.)

substrate in various forms to produce fuel alcohol is a relatively recent development and has yet to prove its economic viability. Lower crude oil values are likely further to delay a build-up of alcohol capacity in China. While the government policy of limiting the use of cereals for ethanol production effectively put a lid on new investments, the low price of oil will act as an additional disincentive. Achieving a consumption target of 2 million tons of fuel alcohol for 2010 looks more than a little doubtful.

India's output of sugar and molasses will be considerably lower in 2008–2009 than it was a year earlier. The downturn has already boosted values of the sugar co-product and as a result, values of alcohol as well. To offset higher domestic molasses prices, ethanol and chemical companies have resorted to importing denatured hydrous alcohol from Brazil. Moreover, arrivals of finished chemical products will also increase as a result of tight supplies of the sugar co-product. The sharp fall in crude oil values is unlikely to change the government's lukewarm stance on biofuels.

The Philippines government remains committed to biofuels. The local alternative fuels sector will grow further despite the decline in world oil prices. The introduction of E-5 blends in 2009 will require 220 to 230 million liters of alcohol in a year. By 2011, an E-10 blend will become mandatory, raising consumption to 480 to 490 million liters per year.

Outlook for Biodiesel Production

After several years of strong growth rates, world biodiesel production is forecast to remain virtually flat in 2009. The outlook strongly depends on the present low fuel prices. On one hand, low energy prices reduce feedstock manufacturing costs. Conversely, they decrease sales values (and thus production margins) for biofuels.

Actual biodiesel consumption figures will strongly rely on the blending demand out-look for conventional fuels as there is currently no real B-100 market. However, the latest IEA data note a decline of conventional fuel consumption. Not only will the expected 2-year contraction in oil demand be the first since the early 1980s, but 2009's decline will also be the largest since 1982 (IEA 2009).

In 2008, EU biodiesel production reached 6 million tonnes (around 45% of global production). The greatest potential for feedstock suppliers inside and outside the EU 27 was the vegetable oil market based on significant import demand from the Community. In particular, consumption of vegetable oil is set to grow if the EU raises its consumption targets after 2010, thereby increasing incentives for oilseed growers.

The spread between average biodiesel ex-works prices and total net production costs narrowed but remained negative. Vegetable oil prices in the first half of 2009 declined slightly more than biodiesel prices. However, the main problem comes from low fuel prices. Lower methanol prices may ease this situation only marginally and only marginal growth of 2009 EU 27 biodiesel output can be expected. Furthermore, competition between imports and EU material may not cease with the end of SME B-99.9. PME is traded at a significant discount against EU specifications and reports of direct SME shipments from Argentina have already appeared. Competitiveness of biodiesel production depends on relative prices of feedstocks and fossil fuels (Figure 7.5).

One recent issue to surface has been the biodiesel trade between the United States. The EU announced an import duty of €20 to 190 per tonne on American biodiesel imports and an anti-subsidy levy of €230 to 260 per tonne for an initial six months. U.S. fuel blends, mainly the so-called SME B-99.9, qualify for a tax credit of $1 per

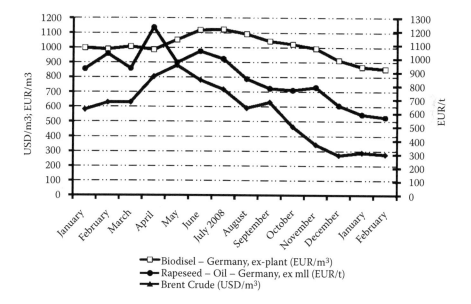

FIGURE 7.5 Prices of biodiesel, crude oil, and rapeseed oil in European Union, January 2008 through February 2009. (*Source:* HGCA, Kingsam and author's calculations.)

gallon, around $300 per tonne, which more than offsets the cost of freight and the Community's import tariff of 6.5%.

Around 1.3 million tonnes of American biodiesel were shipped to the Community in 2008. The growth of the U.S. biodiesel industry was significantly boosted by this business, so the end of this trade is a setback for American manufacturers. Southeast Asian and South American producers may benefit from this development, as some export business previously met by American SME B-99.9 may come their way. As long as freight rates remain low, direct shipments of the fuel are competitive. However, 2009 growth in potential EU biodiesel demand is seen at a mere 4%, driven by a number of increases in blending obligations and expected competition from EU producers (Licht, 2009).

The EU's sustainability requirements may fundamentally change the Community's import demand for biodiesel. Under the Renewable Energy Directive, biofuels must achieve a 35% reduction in GHG emissions from 2010, rising to 50% by 2017. According to the EU's Joint Research Committee figures published in 2008, the use of SME reduces GHG emissions by only 31% while PME without methane capture at the oil mill is even worse at only 19%. That means that after 2010 when the Renewable Energy Directive is to be ratified, export prospects for SME and PME are gloomy. Biodiesel exporters from South America and Southeast Asia as well as the Community's biodiesel producers using these feedstocks may face severe problems after 2010. Significant growth may occur in the use of waste cooking oil and animal fat in the EU, as in both cases GHG reductions stand at 83%. Logistical costs accompany use of these feedstocks (collection and refining of oils) and the feedstock supplies are also limited. Sustainability of SME is an issue in the United States, and the Environmental Protection Agency is currently assessing the national ecological aspects of biofuels.

Hydro treatment and co-processing are technical procedures that have the potential to replace biodiesel. Hydro cracking is a process in which a synthetic fuel is made from biodiesel feedstocks such as animal fat or vegetable oil without esterification. With co-processing, conventional fuel is directly mixed with vegetable oil. Several oil companies such as ConocoPhillips in the U.S. and Finland's Neste Oil have invested significant amounts in plants that already operate although only at modest levels. Considering the sustainability issue, most of these hydro-treated vegetable oils would meet the GHG reduction levels under the Commission's proposal.

All the biofuels produced from wastes, residues, non-food cellulosic materials, and ligno-cellulosic materials shall be considered to be twice the quantities made by other materials. This means that only half the volume of this type of biofuel is needed to achieve the 10% target. However, it does not automatically mean that this biofuel will have a double economic value nor is it certain whether this double counting will offset the higher production costs of most advanced biofuels. It is equally unclear whether higher CO_2 savings will be realized because less volume may result in lower net emission reduction.

Based on the quantitative targets at European and national levels and the EU's present biodiesel manufacturing capacity of about 15 million tonnes, no additional biodiesel plans are needed. On the contrary, European biodiesel manufacturers must make efforts to develop export markets and new sales markets (e.g., for biofuel oil).

At the same time, they should make better use of their advantages in terms of cost and CO_2 balance in a situation where cut-throat international competition is substantially greater. From this perspective, it does not make sense for further subsidies to be provided from Community or national budgets to add more biodiesel capacity.

The end of the SME B-99.9 business also means significantly lower biodiesel output to the United States compared to 2.4 million tonnes in 2008. However, the biodiesel mandate under the Energy Independence and Security Act may help make up for the loss of the Community business. There is no mandate in place for 2009 but it is expected that the 2009 and 2010 mandates will be combined into a two-year period that will create 3.8 million tonnes of consumption. The U.S. federal tax credit expiring Dec. 31, 2009, may be extended. The present prospects for import business appear limited. At current freight rates plus the 4.6% import tariff, Argentine SME is not competitive.

Brazil's B-3 mandate introduced in 2008 raised output to 1 million tonnes and may lead to strong growth in 2009. Brazil's biodiesel production capacity has risen to more than 2.7 million tonnes annually from 1.4 million. The B-3 mandate requires 1.1 million tonnes of biodiesel annually. Most domestic output is destined for domestic use due to the relatively high cost of production. Because of industry overcapacity, a B-4 mandate may be introduced in 2009 according to recent official announcements. It should be noted that 2009 blending demand is expected to be significantly below earlier forecasts, as diesel consumption is set to decline by 10% due to lower economic activity.

Argentina's manufacturers see Europe as their main outlet, and the arrival there of several shipments is expected for 2009 but continue to experience overcapacity as local plants can produce more than 1.5 million tons. Production in 2008 was around 0.8 million tonnes.

Southeast Asian producers expect to benefit from the end of SME B-99.9 based on significant biodiesel import demand from the EU. However, that benefit could be shortlived because of doubts about the sustainability of biodiesel production from palm and soy oils, particularly in the EU. A stop to biodiesel exports to the EU would be a severe setback for Indonesia and Malaysia.

Biofuel Policy Considerations

The growth of biofuels and the impending evolution to second-generation biofuels present considerable challenges in terms of policy development, trade, and certification of sustainability. Such issues were dealt with on a "local" basis earlier, but the time has come to take a global approach. The economics and environmental benefits of first generation biofuels are location-specific. Both the United States and the EU have many of the same players supporting and resisting biofuel growth. The EU appears to be further along in raising issues of sustainability, including mitigating threats to biodiversity, the effect on global warming, and concerns related to food supply. However, these issues are gaining attention on both sides of the Atlantic.

The three traditional biofuels options are bioethanol, biodiesel, and biogas. They differ in terms of feedstock source, net energy yield per hectare, and investment cost. The net energy yield per hectare for biogas can be much higher than with bioethanol production, provided the entire crop is fermented in a biogas plant. However, bio-ethanol approaches the net energy yield of biogas when cellulose is fermented to alcohol. Additionally, the investment costs are much higher for biogas than for bio-ethanol. These differences explain why bioethanol is predominantly produced in countries with abundances of agricultural area such as the U.S. and Brazil.

Ethanol production from corn in the U.S. is totally different from the production of ethanol from sugarcane in Brazil due to the availability of land, energy conversion rates, and technologies used. In more densely populated regions such as the EU, farm land is more expensive. Therefore, the net energy yield per unit area is more impor-tant and so is biogas production. Additionally, the population density causes more waste from food use and livestock production. The more expensive the farmland—and the more waste and manure available—the more attractive the biogas option may become.

The development and evolution of trade rules regarding biofuels is becoming a pivotal issue in both the EU and the U.S. Europe is questioning biofuel production on agricultural lands. The U.S. has more land, but it appears that substantial farmland could be made available in new EU member states. Otherwise, biofuels will have to be supplied by countries outside the EU.

At present, feedstock for biofuel occupies only 1% of crop land. Rising popula-tion, changing diets, and demands for biofuels are estimated to increase demand for crop land by 17% to 44% by 2020. Other evidence indicates that sufficient appropri-ate land is available to meet this demand to 2020, but this must be confirmed before global supplies of biofuel increase significantly. Current policies are not entirely effective in assuring that additional production moves exclusively to suitable areas, and attempts to do so will face challenges in terms of implementation and enforce-ment. Governments should amend but not abandon biofuel policy in an effort to recognize these issues and ensure their policies deliver net GHG benefits.

Strategic Alliances in Biofuel Agribusiness

Strategic alliances have been widely used in the biofuels business, particularly in the research and development phase to seek new sources for production. Oil com-panies have played a significant role in the production and marketing of biofuels worldwide. Their role varies in terms of actual production but the nature of their business requires their involvement in the blending, distribution, transportation, and marketing of biofuels. Brazil's Petrobras, Finland's Neste Oil, and France's Total are examples of vertically integrated energy companies that have enjoyed success in the biofuels marketplace.

Large-scale biofuels production can be managed by large feedstock producers, with little involvement of oil companies. The advent of mandates will serve to hasten strategic alliances and joint ventures of agribusiness and the energy sector. In the future, energy companies will enter agriculture more rapidly, likely causing struc-tural changes that will increase reliance on contracting and vertical integration.

Ethanol producers are price takers, not price makers. Petroleum companies dictate the ethanol price to ethanol producers because of the small number of oil refinery companies and large numbers of ethanol producers. Little information is available to show how ethanol contracts are written and only a few organizations report cash prices. However, the Chicago Board of Trade has established a futures market for ethanol that may provide some transparency about pricing. High oil prices tend to lead to higher biofuel production and delivery costs. If biofuel production significantly increases demand for feedstock, prices for feedstock will be driven upward. The opportunity cost for farmers for raw materials will be tied to the higher-valued market.

Subsidies

The reasons for supporting biofuels are compelling: reducing global warming, enhancing energy security, and promoting rural development. In spite of these benefits, biofuels are critically dependent on policies in the countries that consume them—blocking faster development of markets in developing countries. Mandates and targets for biofuel use combined with government incentives will continue the current emphasis on supply-side economics and focus on internal combustion engines. Total removal of all subsidies and other support for biofuels would reduce the capacity of the industry to respond to the challenges of transforming its supply chain and investing in advanced technologies. On the other hand, the rate of introduction of biofuels should be commensurate with the ability to establish adequate controls and policies.

Advanced technologies required for the production of cellulosic ethanol are immature, expensive, and require incentives to achieve significant market penetration by 2020. However, the biofuels industry promises advancements that will yield cheap, abundant biofuels from all plant materials and wastes. Technologies using genetically modified (GM) plants are being promoted in an effort to streamline processes and reduce costs. Research focuses on GM microbes that can improve the breakdown and fermentation processes to streamline cellulose, reduce lignin, or even change the nature of both materials. Synthetic biology is a new approach involving the use of genetic information to build completely new organisms, albeit with unknown impacts.

Mandatory Blending Targets

Mandatory blending targets serve as a common strategy to establish a foothold for biofuels in the marketplace. As the size of the biofuel market increases, it will be important to verify the expected size of the market and also review the need for increases and dissemination of such targets in light of the following:

- What will decision makers think about biofuels in the near future?
- What decisions must be made about mandatory blending targets?
- What is the long-term influence of these decisions on the competitiveness of the nation?

Additionally, some "uncontrollable" factors to be considered in the world energy sector include:

Political–legal factors — The ratification of the Kyoto Protocol in March 1999 was a turning point for the biofuel market. In this context, ethanol appears as a renewable energy source with the potential of complementing or even replacing some of the fossil fuels in the world's energy matrix, particularly in the transportation sector.

Economic–natural factors — These include the evolution of oil prices and supplies; competition among producers of various forms of renewable energy; the growth in sales of flex-fuel and hybrid vehicles; opening of new markets for ethanol fuel; competition for biomass; and chains of sustainable production, including agricultural practices, human resources, profitability, and access to markets.

Social–cultural factors — These include the growth in numbers of "green" consumers; the positive image of clean fuels; national defense and security; expectations and requirements of corporate social responsibility; concerns about human health (air pollution) and quality of life (improved public transportation); enhancement of fair trade in purchasing decisions; and the perceived value of domestically produced fuels.

Technological factors — The level of investment in research and development should continue to result in advancements including improvement of efficiency in flex-fuel vehicles; new processes for ethanol production, burning biomass, and extraction of cellulosic ethanol; new co-products of biofuel production; and diversification of energy feedstocks and production.

Distiller Grains from Ethanol

Ethanol production structurally changed the economics of the feedgrain–livestock sector as when distiller grains (DGs), a co-product of ethanol production, entered the livestock feed marketplace. As biofuel production grows, the traditional sources of feed grain (mainly corn and feed wheat) will likely take a secondary role to the use of DGs as preferred rations.

DG production helps address the food-and-fuel concerns raised about grain-based ethanol production. The corn traditionally used primarily for livestock feed is now transformed into both fuel and feed—helping satisfy demands in two important areas. We must take into consideration that cellulosic ethanol produces no feed co-product. However, its by-product (primarily lignin) can be used to generate heat. Certain questions about corn ethanol DGs still require attention.

- Can DGs alleviate the projected shortages of grain for feed used by the livestock sector? Current evidence indicates that DGs serve as more than adequate replacements for feed grains, especially as livestock producers incorporate higher levels of DGs into their feeding programs.
- Is it possible to create an international market for DGs? How will issues such as consistency standards, GMOs, and other factors be addressed? It is interesting to note that, in 2008, the U.S., the leading DGs producer, exported some 20% of its total DG output (some 4.51 million tonnes), equivalent in feed value to 4.3 million tonnes of whole corn.
- What will be the global geographic pattern of DG distribution from surplus regions to deficit regions? Ethanol production tends to be concentrated in areas where feedstock and livestock production are abundant. Livestock producers outside these areas will want to avail themselves of this high-value ration.

Trade and Sustainability Certification of Ethanol

A growing number of countries have engaged in the production and use of ethanol as a fuel for transportation—allowing the world's production and exports to double in a short time. A global market of food and biofuel requires the development of expertise in building agribusiness systems that are increasingly transnational and sustainable. This global ethanol market will involve more production, compulsory legislation, and the standardization and certification of the ethanol.

Market structure is influenced by policy, so strengthening the market is essential. Stakeholders focus on their local markets first (the "homegrown" concept is attractive) and international investment in biofuels has been limited. Oil prices are largely demand-driven and global recession has led to significant price drops. Investments in alternative energy sources are risky in this environment without policy measures that ensure against major drops in oil prices. Policy is a key to promote sustainable fuel ethanol trade. At present, uncertain classification, a wide range of government measures (tax incentives, tariffs, subsidies), and a web of varying technical and environmental standards do not facilitate trade.

It should be possible to establish a genuinely sustainable biofuel industry if robust, comprehensive, and mandatory sustainability standards are developed and implemented. The risks of indirect effects can be significantly reduced by ensuring that the production of feedstocks for second-generation biofuels takes place mainly on idle and marginal land—and by encouraging technologies that take best and appropriate advantage of wastes and residues.

Sustainable production is increasingly regarded as a prerequisite for market access. Sustainability certification has three main dimensions: environmental, economic, and social. A schematic for certification must overcome the difficulty inherent in measuring and verifying what, in many cases, are aspirations or principles. Certification requires an institutional environment with requirements that can be effectively and consistently implemented, and an organizational environment that supports reliable monitoring and evaluation.

The main initiative for certification of biofuels has come from national governments, private companies, non-governmental organizations, and international organizations. Most certification proposals are in the early stages, while other may come into force in the near term. They vary considerably in terms of the principles included and the procedures and organizational processes involved. And most are based on existing systems set up for the agriculture, forestry, or energy sectors.

An effective certification system must cover all biomass (regardless of end use) and all relevant bioenergy, and must take a global approach as biomass and bioenergy sources become internationally traded commodities. Systems that focus simply on national or EU-wide implementation, for example, will not help solve major sustainability issues. Additionally, the system must take a holistic approach or risk forfeiting all relevance. For example, if the relatively small quantities of palm oil used for biodiesel are produced in a sustainable manner, but the large volumes consumed in the food sector are not, all the effort expended would be invalid.

As certification criteria are considered, each country should prioritize the areas of law, production and products, communications, distribution and logistics, and human resources. Higher targets for biofuels in the marketplace should be implemented carefully to ensure affected fuels are demonstrably sustainable. Any criterion related to competition or demanding more than a mere reporting obligation could potentially lead to infringement of WTO rules.

ENVIRONMENTAL SECURITY

Biodiversity losses have accelerated, most notably in the tropics. The depletion of fisheries and fish stocks has continued, and in some cases has accelerated. China's growing appetite for mineral and energy resources in Africa and elsewhere is cause for concern, and India, Brazil, South Africa, Angola, and other countries plan to fuel their high growth rates by accelerating resource extraction, and the end of this trend is nowhere in sight.

In terms of climate change and the worldwide ecological situation, the picture is not better—it is a good deal grimmer. By adopting the correct policy mix, we can decouple wealth creation from energy and material consumption just as we decoupled wealth creation from the total number of hours of human labour. That was the great achievement of the industrial revolution. Labor productivity has risen at least 20-fold in the past 150 years of industrialization. Resource productivity should become the core of our next industrial revolution. Technologically speaking, this should be no more difficult than achieving the increase of labor productivity.

We now know that the [over] exploitation of our entire ecosystem and the depletion of natural resources (the reserve-to-production ratio of oil reserves is rapidly declining) carry a price that must be paid today to compensate future generations for the losses (or costs of substitution) they will face tomorrow. Moreover, world population growth by 50% during the next 50 years, causing new scarcities (water) and pollution (CO_2 emission rights), is accelerating these issues. Corporations in energy-intensive sectors must start taking future CO_2 prices into account in their investment decisions and public disclosure policies now. Because the scarcity of emission rights has been recognized, an active market has been created in the EU. CO_2 emission rights now have a price; more regional cap and trade markets for CO_2 have been created in the U.S. and are in the process of development elsewhere.

The environment is now back at center stage after a quarter century of denial among the political and business elites in the U.S. The weight of evidence from the Intergovernmental Panel on Climate Change and the devastating levels of pollution in the industrial centers of the high growth countries like China have at last shifted opinion in favor of tough new controls. The EU has taken the political lead in addressing global warming, setting up the European Trading System (ETS) for CO_2 emissions. President Obama has clearly committed to mitigating global warming, and China too has become serious about tackling pollution, climate change, and energy efficiency. Renewable energy sources now constitute a dynamic growth sector, and the Convention on Biological Diversity (CBD) is enjoying increasing visibility in the signatory states—most countries around the world except the United States.

Never reluctant to waste a good crisis, Joseph Stiglitz and Nicholas Stern made a joint appeal to use the financial crisis as an opportunity to lay the foundations for a new wave of growth based on technologies for a low carbon economy (Financial Times, 2009). The investments would drive growth over the next two or three decades and ensure its sustainability. They noted that "providing a strong, stable carbon price is the single policy action that is likely to have the biggest effect in improving economic efficiency and tackling the climate crisis." Lord Stern calculated that governments should spend at least 20% of their stimulus on green measures to achieve emission targets (Stern, 2006).

The environmental resource scarcity issues are entirely real. As a result of climate changes, most agricultural patterns may become disrupted and the poorest countries are the most vulnerable to such disruptions. Over the long term, environmental security is the mirror image of food security, because we have no food without substantial clean water resources, productive soils, and appropriate climate. In turn, failure to tackle environmental degradation jeopardizes the future of agriculture and land use. Climate change subjects all businesses and society in general to cumulative, long-term risk. The failure of agriculture alone would lead to widespread hunger in developing countries and mass migration of people (half a billion according to the United Nations), mostly to developed countries.

The search for more environmentally friendly agricultural inputs and practices must continue. Scientists are working to improve the efficiency of photosynthesis, carbon capture, nitrogen fixation, and other cellular processes that boost biomass yields. It may also become possible to plant crops in soils lost to salination and genetically modify plants to grow on marginal or otherwise unusable farm lands.

Ecosystems

Mankind is directly influenced by the loss of biodiversity. Through the extinction of species, we lose crucial opportunities to solve many problems of our society. Biodiversity provides us directly with essentials like clean water and air and fertile soil; it protects us from floods and avalanches. These benefits can all be valued economically. It is a difficult and complex task, but such a valuation would clearly show how important biodiversity is for human well-being and economic development (Table 7.1).

Many people are unaware of the speed with which we are consuming our natural resources. We are producing waste far faster than it can be recycled. It is important to compare the needs for public goods and services with arguments whether or not market failures are linked to the provision of services. Market failure is a crucially important justification for taking measures to protect our landscapes. Corrections in market failures may also be achieved through investments and the provision of payments to reward land managers who provide public goods and services (European Commission, 2008).

It is important to demonstrate the economic value of ecosystem goods and services. We must know the costs and also be assured that the benefits are greater than the costs. The consensus about the importance of incorporating these "ecosystem services" into resource management decisions is increasing, but quantifying the levels and values of these services has proven difficult.

Many searches revealed a disappointingly small set of attempts to measure and value these services. The earliest was the quantification of global ecosystem services by Constanza et al. (1997). Estimates of values based on willingness to pay

TABLE 7.1
Future Environmental Scenario to 2050

Use	2000 Million km²	2010 Million km²	2050 Million km²	Difference 2000 to 2010	Difference 2010 to 2050	Difference 2000 to 2050
Natural areas	65.5	62.8	58.0	–4%	–8%	–11%
Bare natural areas	3.3	3.1	3.0	–6%	–4%	–9%
Managed forests	4.2	4.4	7.0	5%	62%	70%
Extensive agriculture	5.0	4.5	3.0	–9%	–33%	–39%
Intensive agriculture	11.0	12.9	15.8	17%	23%	44%
Woody biofuels	0.1	0.1	0.5	35%	437%	626%
Cultivated grazing	19.1	20.3	20.8	6%	2%	9%
Artificial surfaces	0.2	0.2	0.2	0%	0%	0%
World Total	108.4	108.4	108.4	0%	0%	0%

Source: Braat, L., and Brink, ten P., Eds. 2008. *Contribution of Different Pressures to the Global Biodiversity Loss between 2000 and 2050 in the OECD Baseline*: Interim Report. Brussels: The Economics of Ecosystems and Biodiversity (TEEB).

for a hectare's worth of each of the services were extracted from the literature and expressed in 1994 U.S. dollars per hectare. Some attempt was made to adjust the values across regions by purchasing power. The central estimate of the total value of annual global flows of ecosystem services in the mid 1990s was $33 trillion; the range was thought to be $16 trillion to $54 trillion. To put their figure into some kind of context, the central estimate was 1.8 times larger than the global gross domestic product (GDP) at that time. We should take the figures only as the roughest approximation. The authors noted the huge uncertainties involved in making calculations of this kind.

Another study titled "Millennium Ecosystem Assessment" (MA), found that over the second half of the 20th century human capacity to exploit ecosystems increased dramatically to meet rapidly growing demands for food, fresh water, timber, fiber, and fuel. This exploitation has resulted in a substantial and largely irreversible loss of biodiversity of life. The benefits of these developments have been unevenly distributed and have caused uncomfortable tradeoffs among the services provided by ecosystems (United Nations, 2003).

The findings of "The Ecosystems and Human Well-being–Biodiversity Synthesis" report established the importance of biodiversity of environmental or ecosystem services to human well-being (Reid et al., 2005). The report is based on the findings of the MA and supports the goals of improving the management of the world's ecosystems, providing better information to policy makers, and building human and institutional capacity to conduct integrated assessments. The challenge of sustainably managing ecosystems for human well-being must be met through institutions at multiple scales—no individual scale is critical. Local, national, regional, and international institutions have unique roles to play in understanding and managing ecosystems for people. Ecosystems provide many tangible benefits ("ecosystem services") to people around the world.

The Stern Review (2006) parallels the TEEB report of the European Communities study of the economics of climate change (2008). Climate change will exert very serious impacts on growth and development. The costs of stabilizing the climate are significant but manageable; delay would be dangerous and far more costly. The review estimates that if we fail to act, the the costs and risks of climate change will be equivalent to losing at least 5% of global GDP each year, now and forever. In contrast, the costs of action—reducing GHG emissions to avoid the worst impacts of climate change—can be limited to around 1% of global GDP each year. Key to understanding the conclusions is that as forests decline, nature stops providing services it provided essentially at no cost. Thus the human economy must step in to provide them, for example, by building reservoirs, building facilities to sequester carbon dioxide, and farming foods that were once naturally available.

The "Living Planet Report" of the World Wildlife Fund (WWF, 2008) demonstrates that humans live far beyond the capacity of the environment to supply us with services and absorb our waste, and expresses the concepts of ecological footprints and biocapacity as hectares per person.* Humanity's footprint first exceeded global biocapacity in 1980 and the overshoot has increased regularly since then. In 2005, the WWF calculated the average global footprint across the world was 2.7 global hectares (gha) per person[†] compared to a biocapacity calculated as 2.1 gha/person— a difference of 30%. That means each person on earth consumes on average 30% more resources and waste absorption capacity than the world can provide. We are destroying the earth's capacity and compromising future generations.

The TEEB study (2008) of "The Economics of Ecosystems and Biodiversity" concerns the struggle to find the value of nature. At present, about 100,000 terrestrial protected areas cover 11% of the earth's land mass. These protected areas provide ecosystem services and biodiversity benefits valued at $4.4 trillion to $5.2 trillion (a million millions) per annum. If you want a comparison, that amount is more than the revenues of the global automobile, steel, and IT services sectors of the economy combined! Calculations show that the global economy is losing more money from the disappearance of forests than through the current banking crisis. Forest decline costs about 7% of global GDP, putting the annual cost of forest loss between $2 trillion and $5 trillion, based on adding the values of the various services that forests perform, (providing clean water and absorbing carbon dioxide). However, the cost falls disproportionately on the poor, because more of their livelihoods depends directly on forests, especially in tropical regions. The greatest cost to Western nations would initially come from loss of a natural absorber of the most important GHG (European Commission, 2008).

The study shows that diversity is crucial for survival and stresses the importance of biodiversity for economic development. It may be possible to substitute some

---

* The ecological footprint measures the "biologically productive land and water area required to produce the resources an individual, population or activity consumes and to absorb the waste it generates, given prevailing technology and resource management" (WWF, 2008).
[†] A global hectare has a global average ability to produce resources and absorb wastes.

ecosystem services by human-made technologies, but the study results clearly show that it is often cheaper to invest in conserving biodiversity than paying for new technologies to replace the services nature provides. Therefore it is essential for the safeguarding of our natural resources to coordinate economic interests. We must give the ecosystem services of biodiversity a market value to create incentives for developing countries to conserve their biodiversity.

Market-based instruments are helpful for providing the world a chance to secure natural resources and secure individual livelihoods simultaneously. The inclusion of the private sector into the processes of conservation and sustainable use of biodiversity has high priority. The goals of conservation and sustainability will be achieved only if the main drivers of ecosystem and biodiversity loss are addressed through appropriate intervention and response based on credible valuations. Businesses must accept biodiversity as an indispensable resource (which it is) and treat this resource with respect and care. The aim of the initiation of TEEB program and the engagement of the private sector is ultimately to improve decision making. The conservation of biodiversity is supported by a number of good economical, ethical, and spiritual reasons that should give it a high priority in decision making. Realizing the economic value of biodiversity can provide us with additional irresistible arguments for the conservation and sustainable use of biological diversity for future generations.

The Global Canopy Programme's report concludes: "If we lose forests, we lose the fight against climate change." International demand has driven intensive agriculture, logging, and ranching, leading to deforestation. Standing forest was not included in the original Kyoto Protocol and stands outside the carbon markets. The inclusion of standing forests in internationally regulated carbon markets could provide cash incentives to halt this disastrous process. Marketing these ecosystem services could provide the added value forests need and help dampen the effects of industrial emissions. Countries wise enough to have kept their forests may find themselves the owners of a new billion dollar industry (Parker et al., 2008).

Currently, two paradigms for generating ecosystem service assessments are meant to influence policy decisions. Under the first, researchers use broad-scale assessments of multiple services to extrapolate a few estimates of values, based on habitat types, to entire regions or the entire planet (Costanza et al., 1997). This "benefits transfer" approach incorrectly assumes that every hectare of a given habitat type is of equal value—regardless of its quality, rarity, spatial configuration, size, proximity to population centers, or prevailing social practices and values. Furthermore, this approach does not allow for analyses of service provision and changes in value under new conditions.

In contrast, under the second paradigm for generating policy-relevant ecosystem service assessments, researchers carefully model the production of a single service in a small area with an "ecological production function" based on how provision of that service depends on local ecological variables (Kaiser and Roumasset, 2002; Ricketts et al., 2004). These methods lack both the scope (number of services) and scale (geographic and temporal) to be relevant for most policy questions (Nelson et al., 2009).

Spatially explicit values of services across landscapes that may inform land use and management decisions are still lacking. Quantifying ecosystem services in a spatially explicit manner and analyzing tradeoffs between them can help to make natural resource decisions more effective, efficient, and defensible (Nelson, et al., 2009). Both the costs and the benefits of biodiversity-enhancing land-use measures are subject to spatial variation, and the criterion of cost effectiveness calls for spatially heterogeneous compensation payments (Drechsler and Waetzold, 2005). Cost effectiveness may also be achieved by paying compensation for results rather than measures. We must ensure that all the possibilities to create markets to provide environmental services are fully exploited to minimize the public costs (and the involvement of government bureaucracy).

Creating markets for environmental services may encourage the adoption of farming practices that provide cleaner air and water and other conservation benefits. Products expected to generate the greatest net returns are generally selected for production. Since environmental services generally do not have markets, they have little or no value when farmers make land use or production decisions. As a result, environmental services are under-provided by farmers. The biggest reason that markets for environmental services do not develop naturally is that the characteristics of the services defy ownership. After environmental services are produced, people can "consume" them without paying a price. Most consumers are unwilling to pay for a product they formerly obtained for nothing. This prevents markets from developing. Can anything be done other than relying on government programs to provide publicly funded investments in environmental services?

Creating markets for environmental services is not an entirely novel idea. Governments play a central role in setting them up, as has been done already for water quality trading, carbon trading, and wetland damage mitigation. These markets would not exist without government programs requiring regulated business firms (such as industrial plants and land developers) to meet strict environmental standards. In essence, legally binding caps on emissions (water and carbon) or mandatory replacement of lost biodiversity (wetland damage mitigation) create the demand needed to support a market for environmental services. So-called cap-and-trade programs create a tradable good related to an environmental service (Ribaudo et al., 2008).

Mandatory reduction pledges can be obtained from all developed nations. The same principle can apply to project-level reductions in developing countries. Mandatory cap-and-trade programs have been introduced in the northeastern U.S. and the EU. The U.S. and Australian governments will also institute mandatory cap-and-trade programs to create financial incentives to limit energy use and reduce emissions.

In the case of water quality, it is necessary to establish caps on total pollutant discharges from regulated firms in some watersheds and issue discharge allowances specifying how much pollution each firm can legally discharge. In markets for greenhouse gases, carbon credits are exchanged. Contracts also include renewable energy credits and voluntary carbon credits.

No-net-loss requirements for new housing and commercial development require replacement of damaged or lost wetland services, creating demands for mitigation credits obtained by creating new wetlands. In all these cases, the managing or regulatory entity defines the tradable good and enforces all transactions.

Simply creating demand for an environmental service does not guarantee that a market for services from agricultural sources will actually develop. A number of impediments affect agricultural producers' ability to participate in markets for environmental services. Purchasers may be unwilling to enter a contract with a farmer who cannot guarantee delivery of the agreed-upon quantity of pollution abatement, wetlands service, or other environmental commodity. Some markets prevent the selling of uncertain services. For example, the Chicago Climate Exchange does not certify credits from soil types for which scientific evidence about their ability to sequester carbon is lacking. Transaction costs can also undermine the development of markets for environmental services (Ribaudo et al., 2008).

If markets are to become important tools for generating resources for conservation on farm lands, governments and other organizations may have to help emerging markets overcome uncertainty and transaction costs. Governments can reduce uncertainty by setting standards for environmental services and play a major role in reducing uncertainty by providing research about levels of environmental services from different conservation practices. For example, the government can develop an online nitrogen trading tool to help farmers determine how many potential nitrogen credits their farms can generate for sale in a water quality trading program.

While markets have many desirable properties, what they can acomplish is limited, even with government assistance. Public-good characteristics that defy ownership discourage markets for environmental services from developing and prevent the reflections of the full values of environmental services in prices The prices of credits in water, carbon, and wetland markets also may not reflect their full social value and indicate only their value to the regulated community. A national cap-and-trade program could establish a national market for carbon credits. Other markets such as water quality trading or wetland damage or loss mitigation may be limited to a few specific geographic areas.

Appropriate land and environment will exert significant roles on EU policy and budgeting. The EU needs regulation defining its policy on markets for environmental services. This policy would cooperate with member states (MS) and local governments to establish a role for agriculture in environmental markets. We must find ways to make EU policies and programs support producers that want to participate in such markets. Conducting research and developing tools for quantifying environmental impacts of farming practices is of great importance as well. Technical guidelines must be established to measure environmental services from conservation and other land management activities, with priority for participation in carbon markets. Guidelines are also needed for establishment of a registry to record and maintain information about environmental service benefits and a process for verifying implementation of conservation or land management activities reported in the registry. Enthusiasm can be observed for green public procurement linked to certification and labeling and supported by data on embedded water, carbon, and biodiversity along with guidance to help public procurers buy less biodiversity-harming commodities. All these measures are useful stepping stones toward biodiversity-reflective procurement by public sector establishments such as schools and hospitals.

Ecosystems markets will change the present, economics-only value paradigm that requires winners and losers. As an example, countries and companies with

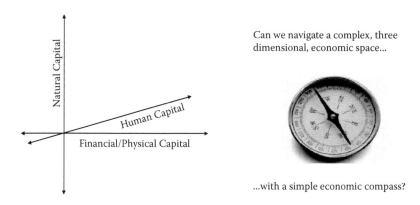

FIGURE 7.6   The Economics of Ecosystems and Biodiversity (TEEB): navigation challenge ahead. (*Source:* European Commission, 2008.)

significant carbon-sink potential would benefit. Conversely applying the polluter-must-pay principle, CO_2 emitters would pay to continue their emitting activities. The concept of limiting (capping), auctioning, and trading emission, access, and user rights must be further developed beyond CO_2, to include water and other resources on a worldwide scale. Valuing our ecosystems and regulating the access thereto will create a market for payment for ecosystem access entitlements and services. We must upgrade our performance metrics. The values of human and social capital, education, culture, social cohesion, and other factors should be established and more prominently involved in investment and development decisions (Figure 7.6).

Common Agriculture Policy

The Common Agriculturel Policy (CAP) is at a crossroads. While globalization is changing the roles of rural communitie, CAP reform, budgetary constraints, and WTO negotiations affect the support of agricultural communities. The big 2003 reform of the CAP started a process of staged reviews. The decoupling of aid became a frame of reference for several reforms of agricultural sectors. The EU budget review must become more coherent with a new, more effective CAP adapted to the needs of expansion and globalization (its two main challenges in the medium and long terms).

The next reform of the CAP must respond to the new challenges it faces in the first half of this century These include the demand for public goods and services, such as ecosystem services including carbon sequestration and the preservation of less favored areas that farmers provide and are not paid for by the market. This subject is attracting growing attention at a time when the trend in agricultural support points downward as temperatures, emissions, and environmental degradation trend upward. If farm prices rise in the long term, presumably along with production costs, the opportunity costs of land managers to produce ecosystems and other public goods and services will rise as well. Current market signals are distorted because scarce resources such as clean atmosphere, biodiversity, and beautiful landscapes have been priced at zero. Resource degradation is the consequence.

In general, CAP reform is perceived in the EU as driven by the larger member states that account for most agricultural production in the EU. Under current arrangements, several member states receive lower rates of direct support per hectare under Pillar 1 of the CAP than others. It is commonly thought that disparities in CAP support rates among member states have contributed to the cessation of agriculture in more marginal regions and subsequent concentration of production in areas of Europe where CAP payments are higher. The perception is that this practice disregards the needs of smaller member states that have less competitive agricultural sectors (Herzon, 2008).

In reality, no single European rural development policy exists. The various national and regional schemes now in place have been carefully devised for 2007 through 2013 and focused to address [perceived] local priorities. As an example, the proportion of Pillar 2 programs dedicated to Axis 2 (agri-environment and land management measures) ranges from 80% in Ireland to barely 30% in the Netherlands. Spending on Axis 3 (broader rural regeneration policies) ranges from 5% of the national total in Denmark to well over 20% in Germany, the Netherlands, and some of the new member states. These carefully crafted seven-year schemes must be rewritten to account for the Health Check agreement on modulation. The progressive increase toward a 10% modulation rate by 2012 will create an estimated additional €3.241 billion to spend on rural development for 2010 through 2013, but the money comes with strings attached.

CAP reform is likely to depend both on the outcomes of the EU budget review in terms of funding available for the agricultural sector in the future and on other factors such as climate, energy prices, food production, and changing economic conditions that in combination serve to shape the priorities of the citizens of Europe. The traditional CAP objective of food security will remain in place. In the future, agricultural policy must respond to public demands linked to the maintenance of landscapes, conservation of natural resources and biodiversity, food safety, and sustainability.

European society is becoming increasingly urban and people in rural areas are at risk of becoming social minorities with reduced political and electoral clout. However, many of those living in rural areas are responsible for managing the land and sustaining our most valuable natural resources: water, soil, and wildlife biodiversity. Food is more than a commodity. The major challenge for the food system—and thus for farmers—is to become visible and earn the recognition of all European consumers based on its quality, safety, and product diversity.

The EU has an approximately balanced trade account in agricultural goods with third countries. The current decoupled income aid suffers from substantial legitimacy problems because it establishes the various levels of support based on earlier sectors and territories. Furthermore, it does not reflect the changes in the orientations of farms after decoupling after flexibility to produce is established, and this distorts the markets. The Health Check is introducing the gradual step of a decoupling model based on historical production references to a territorial support model. The formalization of a territorial support model within the first pillar of the CAP has the additional effect of gradual material convergence (in terms of nature and functionality) of decoupled payments per hectare and a certain portion of the aid currently granted under the second pillar.

In contrast, MS may make full use of the expanded options available under the Health Check rules for channelling additional support for their fragile dairy, beef, and sheep sectors (Article 68). This offers member states the right to fund tailor-made support by top slicing up to 10% from existing single farm payment (SFP) entitlements and recycling them to specifically targeted aid schemes, representing a partial restoration of the link of production and subsidy (in the dairy, beef, and sheep sectors). MS have to respect the maximum coupled ceiling of 3.5% within the overall of 10% limit.

In 2006, the budget of the rural development policy was in fact reduced, in both absolute and percentage terms. The only tool the EU has to remedy this situation is increasing modulation. The axes for action on rural development are constantly extended, thus invalidating any additional financial effort. The use of modulation is also invalidated by the lack of connection between the resources available to each member state and the extent of its rural problems. Modulation is a percentage of each member state's direct aid. This is beneficial to some countries but condemns smaller states with lower levels of support to merely symbolic modulation.

To avoid huge cuts on the spending side, a strong focus has to be put on delivering European public goods. In order to strengthen multifunctionality, present rural development and agri-environmental payments must be linked more closely to genuinely European public goods. Agricultural payments must therefore be justified by the public goods agriculture can provide by providing services of public interest to the whole of society.

The European Commission is about to submit a white paper on "Adapting to Climate Change" along with a working document. It apparently advocates strengthening the CAP to discourage unsustainable practices and assesses the needs for further regulation. This should help enhance efficient agricultural water use, increase water storage capacity of ecosystems, and reduce flood risks. The white paper indicates the need for financial support for investments aimed at facilitating adaptation to climate change.

REFERENCES

Braat, L., and ten Brinks, P., Eds. *Contribution of Different Pressures to the Global Biodiversity Loss between 2000 and 2050 in the OECD Baseline: Interim Report.* Brussels: The Economics of Ecosystems and Biodiversity (TEEB).

Costanza, R. et al., 1997. The value of the world's ecosystem services and natural capital. *Nature* 387: 253–60.

Drechsler, M. and F. Waetzold. 2005. Spatially uniform versus spatially heterogeneous compensation payments for biodiversity-enhancing land-use measures. *Environ. Res. Econ.* 31: 73–93.

European Communities. 2008. *The Economics of Ecosystems and Biodiversity: Interim Report.* Brussels: TEEB. http://ec.europa.eu/environment/nature/biodiversity/economics/index_en.htm.

FAO. 2003. *World Agriculture: Towards 2015/2030.* Rome: Food and Agriculture Organization of the United Nations.

FAO. 2008. *The State of Food Insecurity in the World.* Rome: Food and Agriculture Organization of the United Nations.

Financial Times. 2009. Obama's chance to lead the green recovery. March 3.

HGCA. 2009. The Agricultural and Horticultural Development Board (AHDB). Subdivision GGCA. London, UK. http://www.hgca.com/content.output.

Herzon, I. 2008. *CAP Reform Profile: Finland*, http://cap2020.ieep.eu/member-states/finland.

IEA. 2008. *World Energy Outlook 2008*. Paris: International Energy Agency.

IFPRI. 2008. *International Food Prices: The What, Who, and How of Proposed Policy Action.* Washington: International Food Policy Research Institute.

IWMI. 2007. *Water for Food, Water for Life: A Comprehensive Assessment of Water Management in Agriculture.* London: Earthscan; Colombo: International Water Management Institute.

Kaiser, B. and J. Roumasset 2002. Valuing indirect ecosystem services: The case of tropical watersheds. *Environ. Dev. Econ.* 7: 701–714.

Krugman, P. 2009. Is a new architecture required for financing food and environmental security? Summary of speech made at launch of Second Forum for the Future of Agriculture, Brussels. http://www.elo.org (accessed April 15, 2009).

Licht, F.O. 2009. *World Ethanol and Biofuel Report (Jan.–Apr.)*. London: Agra Informa.

Nelson, E. et al. 2009. Modelling multiple ecosystem services, biodiversity conservation, commodity production, and tradeoffs at landscape scales. *Front Ecol. Environ.* 7: 4–11.

OECD 2007. *Biofuels: Is the Cure Worse than the Disease?* Paris: Organisation for Economic Co-operation and Development.

Parker, C. et al. 2008. *The Little Reed Book: Reducing Emissions from Deforestation and (Forest) Degradation Global Canopy Programme.* Oxford: John Krebs Field Station.

Reid, W.V. et al. 2005. *Millennium Ecosystem Assessment: Ecosystems and Human Well-being, Biodiversity Synthesis.* Washington: World Resources Institute.

RFA, 2008. *The Gallagher Review of Indirect Effects of Biofuels Production.* London: Renewable Fuels Agency.

Ribaudo, M. et al. 2008. *The Use of Market to Increase Private Investment in Environmental Stewardship.* Washington: USDA ERS.

Stern, N. 2006: *Stern Review: The Economics of Climate Change.* Cambridge: Cambridge University Press.

The Earth Institute. 2005. *The Growing Urbanization of the World.* New York: Columbia University.

United Nations. 2003. *Millennium Ecosystem Assessment: Ecosystems and Human Well-being.* Washington: Island Press.

USDA 2008. *Agricultural Projections to 2017.* Washington: U.S. Department of Agriculture.

WWF. 2008. *Living Planet Report*, Gland, Switzerland: World Wildlife Fund.

8 Opportunities and Challenges of Biofuel Production

Rattan Lal

CONTENTS

INTRODUCTION

In his inaugural address, President Obama declared "We will harness the sun and the winds and the soil to fuel our cars and run our factories." The word "soil" in his address refers to the production of biofuels. The idea of producing liquid biofuels from biomass goes back to 1925 when Henry Ford predicted, "The fuel of the future … is going to come from … every bit of vegetable matter that can be fermented" (*New York Times,* Sept. 24, 1925; Sticklen, 2007). Indeed, Henry Ford's first car, the 1896 Quadracycle, ran on pure ethanol, and the first gas station to sell 10% ethanol blend in the U.S. opened in 1933 (Sticklen, 2007).

The strong emphasis on biofuels in the 21st century is attributed primarily to four interrelated factors (Figure 8.1): (1) increasing atmospheric concentrations of CO_2 and other greenhouse gases (GHGs), (2) risks of abrupt climate change, (3) rapidly increasing global energy demand, and (4) food insecurity affecting more than 1 billion people or about 15% of the global population. Production of biofuels can also improve the rural economy by creation of new jobs. Biomass-based electricity schemes provide more than 9 GWe of generation capacity worldwide (Mirza et al., 2008). Biofuels can create a ripple effect via the linkage of soil quality, food security,

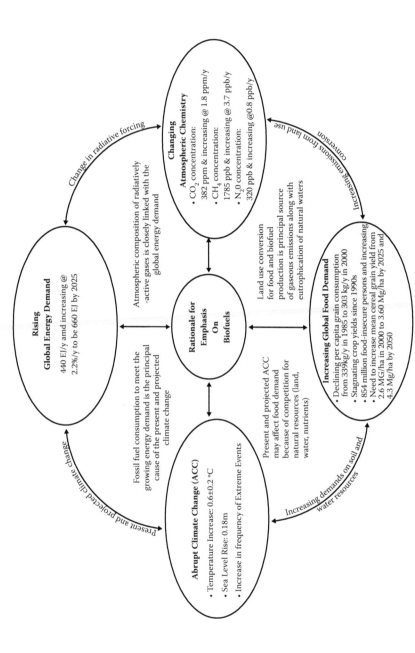

FIGURE 8.1 Four mutually reinforcing factors responsible for increasing emphasis on biofuels. (Data from Wild, 2003; IPCC, 2007; EIA, 2007; Borlaug, 2007.)

TABLE 8.1

World Total Energy Consumption

Region	Energy Demand (EJ/Y)				
	1990	2003	2010	2020	2030
North America	106	123	138	156	175
Europe	144	137	147	165	182
Asia	78	125	175	228	285
Middle East	12	20	26	33	40
Africa	10	14	19	23	28
Central/South America	15	23	30	38	48
World	365	442	535	643	758

Source: EIA. 2006. International Energy Outlook. Energy Information Administration, U.S. Department of Energy, Washington.

and the environment (Naylor et al., 2007). Because of the strong interaction, these factors mutually reinforce one another and create positive feedbacks. Biofuels that derive their energy from biomass and thus serve as carbon (C) neutral energy sources are of interest because they exert little or no effect on the atmospheric concentrations of radiatively-active gases.

Global energy demand was 365 EJ in 1990, 418 EJ in 2001, and 442 EJ in 2003. It is projected to increase to 535 EJ in 2010, 643 EJ in 2020, 758 EJ in 2030, and ~1,000 EJ in 2050 (Table 8.1). Energy demand is rapidly increasing in Asia and other emerging economies. The principal source of global energy is fossil fuel combustion— also a major cause of the enrichment of atmospheric concentrations of GHGs. Total CO_2–C emissions from the U.S. increased from 1,677 Mt C/year (Mt = 1 million metric tonnes = 10^{12} g = teragram) in 1990 to 1,924 Mt C/year in 2006 (Table 8.2). Considering the land-based C sinks, the net emissions from the U.S. increased from

TABLE 8.2

Trends in U.S. Emission of CO_2–C

Year	Total Emissions (Mt C/Year)	Sink (Mt C/Year)	Net Emissions (Mt C/Year)
1990	1,677	201	1,476
1995	1,771	211	1,560
2000	1,918	184	1734
2005	1,944	240	1,704
2006	1,924	241	1,683

Mt = million metric tonnes or teragrams.

Source: Recalculated from EPA. 2008. Inventory of U.S. Greenhouse Gas Emissions and Sinks: 1990–2006.

TABLE 8.3

Sources of Global Emission of CO_2–C Equivalent in 2000

Sources	Total (Gt CE/Year)	% of Total
Energy Related Emissions		
Power	2.75	24
Transport	1.61	14
Industry	1.61	14
Building	0.92	8
Other Energy	0.58	5
Non-Energy Emissions		
Land Use	2.07	18
Agriculture	1.61	14
Waste	0.35	3
Total	11.5	100

Source: Recalculated from Koonin, S.E. 2008. The challenge of CO_2 stabilization. *Elements* 4: 294–295.

1,476 Mt C/year in 1990 to 1,683 Mt C/year in 2006 (Table 8.2). Global CO_2–C emissions in 2000 were estimated at 11.5 Gt C/year (Gt = gigaton = 1 billion metric tonnes = 10^{15} g = petagram) (Koonin, 2008). Energy-related emissions accounted for 65% of the total (Table 8.2). Principal sources of non-energy emissions are land use and agriculture that together account for emissions of 3.68 Gt CE/year (Table 8.3). In addition to CO_2, gaseous emissions from waste (landfill) and non-energy sources also include methane (CH_4) and nitrous oxide (N_2O), with global warming potentials (GWPs) of 21 and 310, respectively (IPCC, 2007a and b). A strong and direct relationship exists between increased energy demand and increased gaseous emissions.

Energy demand and gaseous emissions are also linked with rising food demand because of the increase in world population. There are 1,020 billion food-insecure people in the world (FAO, 2009), and it is widely feared that the United Nations Millennium Development Goals will not be met. The recent increases in food staple prices worldwide has also exacerbated global food insecurity (Trostle, 2008). Global food grain production increased by a factor of 3 during the second half of the 20th century, from 650 Mt in 1950 to 1900 Mt in 2000. However, the mean cereal grain yield of 2.64 t/ha (t = 1 metric tonnes = 10^6 g = 1 Mg) in 2000 must be increased to 3.60 t/ha by 2025 (+35%) and 4.30 t/ha by 2050 (+50%). With probable changes in dietary preferences in Asia and elsewhere in emerging economies, average cereal grain yield must be increased to 4.40 t/ha (+ 62%) by 2025 and 6.0 t/ha (+ 121%) by 2050 (Wild, 2003). The increasing demand for food production will accentuate emissions of GHGs from non-energy sources (e.g., land use conversion and agricultural activities).

Intense debate surrounds the effectiveness of replacing some of the fossil fuel use, nationally and internationally, by biofuels. If biofuels are C neutral as perceived, they can provide renewable energy without drastically affecting atmospheric concentrations of CO_2 and other GHGs. However, serious questions concern viable sources of biofuel feedstock, competition for natural resources (e.g., land, water, and nutrients) to produce the biomass, and energy output:input ratios. This chapter discusses the potentials and challenges of producing cellulosic biofuels, especially in relation to their impacts on soil, water, and the ecosystem C budget.

TRADITIONAL BIOFUELS

Traditional biofuels—biomass used directly for household energy, especially for cooking—have been used since the dawn of human civilization. Principal sources of household energy include woody biomass, crop residues, animal dung, and other agricultural and forestry by-products. Traditional biofuels remain major sources of household energy throughout the developing world. Yevich and Logan (2003) estimated that of ~2 Gt of crop residues available in developing countries of Asia, Africa, and Latin America, 408 Mt (20%) served as traditional biofuels. Venkatraman et al. (2005) estimated that annual global consumption of traditional biofuels included 1,324 Mt of fuel wood, 136 Mt of dried cattle manure, and 597 Mt of crop residues (Table 8.4).

Traditional biofuels are important sources of household energy in India (Pachauri and Spreng, 2002; Pachauri, 2004; Reddy, 2003; WEO, 2002). In Pakistan, an average household uses 2,325 kg of firewood, or 1,480 kg of dung, or 1,160 kg of crop residues annually as cooking fuel (Mirza et al., 2008). Alternatives to traditional fuels based on the use of crop residues and animal dung are being developed (Ravindranath, 1997) because of rapid increases in

TABLE 8.4
Estimates of Traditional Biofuel Use for Residential Energy

		Biofuel Consumption (Tg/Year)			
Region	Base Year	Fuel Wood	Dried Cattle Manure	Crop Residues	Total
India	1995	281	62	36	379
Asia		865	165	488	1,518
World		1,470	280	575	2,325
India	1985	220	93	86	399
Asia		753	133	545	1,431
World		1,324	136	597	2,057

Source: Adapted from Venkatraman, C. et al. 2005. Residential biofuels in South Asia: Carbonaceous aerosol emissions and climate impacts. *Science* 307: 1454–1456. (Tg = 10^{12} g.)

energy demand (Mathur et al., 2004; Naidu, 1996; Suganthi and Williams, 2000). However, indiscriminate removal of crop residues and animal wastes exacerbates the already severe problems in India (and elsewhere in developing countries) of accelerated soil erosion and siltation (Singh et al., 1992; Suresh et al., 2000), depletion of soil organic matter (SOM) and plant nutrients (Muir, 2001), and emissions of soot (black carbon), aerosols, and other pollutants to the atmosphere (Habib et al., 2004).

In addition to CO_2 emissions that are also increasing (Srivastava, 1997), traditional biofuels constitute a serious environmental hazard. Venkataraman et al. (2005) concluded that traditional biofuel combustion is the largest source of black carbon (soot) emission estimated at 160 to 172 Tt/year (Tt = thousand metric tonnes = 10^9 g = 1 gigagram = 1 Gg) in India, 635 to 775 Tt/year elsewhere in Asia, and 920 to 1,230 Tt/year in the rest of the world. New methodologies for estimating biofuel consumption for cooking and emissions of black C and SO_2 have also been proposed (Habib et al., 2004). The aerosols emitted by combustion of traditional biofuels can alter the amount of solar irradiance reaching earth's surface and affect monsoonal rains and the hydrologic cycle (Ramanathan et al., 2001).

The so-called brown clouds over South Asia constitute a major environmental concern (Gustafsson et al., 2009). Jacobson (2002) proposed that reducing emissions of black C may be an effective strategy for slowing global warming. Combustion of dried animal dung and use of traditional biofuels under unventilated conditions are principal causes of indoor pollution (Zhang and Smith, 2005; Mudway et al., 2005; Dasgupta et al., 2009; Kumie et al., 2009). Balilis et al. (2005) observed that under business-as-usual scenarios, household indoor pollution will cause an estimated 9.8 million premature deaths in Africa by 2030.

Use of traditional biomass energy (crop residues and animal manure) severely affects soil quality. Depletion of the SOM pool and the attendant decline in soil fertility and structural attributes (because crop residues and animal manure are not used as soil amendments) exert severe adverse impacts on agronomic productivity. It is estimated that animal dung used as a household fuel in India would be worth U.S. $800 million annually if used as fertilizer (WEO, 2002).

Several beneficial impacts arise from retention of crop residues as mulch on the soil surface (Table 8.5). The most important is an increase in agronomic yields based on soil and water conservation, moderation of soil temperature, increase in activity and species diversity of soil fauna, and improvements in soil structure. Similar benefits are realized through application of animal manure as compost. In contrast, removal of crop residues produces numerous adverse effects on soil quality and the environment. One major adverse impact is the increased risk of soil erosion, nonpoint source pollution, and erosion-induced decline in soil productivity. Residue removal reduces soil resilience and aggravates soil degradation and desertification.

MODERN BIOFUELS

Biomass can be used to generate electricity and heat but can also be converted into liquid fuels for transport, such as methanol, ethanol, and diesel (Figure 8.2). Modern biofuels involve conversion of biomass into liquid (bioethanol, biodiesel) or

TABLE 8.5

Estimates of U.S. Bioethanol and Biodiesel Production

Year	Bioethanol (Billion Liters)	Biodiesel (Million Liters)
1981	0.31	–
1985	2.34	–
1990	2.83	–
1995	5.23	–
2000	6.26	–
2001	6.60	34.10
2002	7.85	37.85
2003	10.70	53.00
2004	13.44	106.00
2005	15.36	344.44
2006	20.75	946.25
2007	25.91	1,858.44

Source: Recalculated from EIA. 2006. International Energy Outlook. Energy Information Administration, U.S. Department of Energy, Washington; EIA. 2007. Annual Energy Review. Energy Information Administration, U.S. Department of Energy, Washington; www.eia.gov/iea.

gaseous (biogas) fuels rather than direct combustion in traditional or modern settings (Ohlrogge et al., 2009). Processes by which biomass is converted into modern biofuels include pyrolysis, gasification, hydrogasification, liquefaction, anaerobic digestion, fermentation, and transesterification (Figure 8.3). In addition, co-generation technology allows direct electricity production and there are numerous pathways of converting biomass into hydrocarbons (Regalbuto, 2009; Williams, 2009).

Worldwide energy consumption increased 17-fold during the 20th century (Demribas, 2007). Proven worldwide reserves of fossil fuels in 2004 were estimated at 1.27 trillion barrels of oil and 6,100 trillion cubic feet of gas (172×10^6 m^3). At the present consumption level of 85 million barrels per day of oil and 260 billion cubic feet (7.36×10^6 m^3) per day of natural gas, the proven reserves represent 40 years of oil and 64 years of natural gas (Vasudevan and Briggs, 2008). The concerns about energy scarcity have led to increased replacement of fossil fuels by modern biofuels. Biofuel production is also increasing globally, and strong supporters of green energy consider biofuel as a viable alternative to reducing dependence on fossil fuel (Goldemberg, 2007; Mathews, 2007; Goldemberg and Guardabassi, 2009).

Global ethanol production has steadily increased from 20 billion liters (BL) in 1997 to 50 BL in 2007. U.S. ethanol production increased from 6 BL in 1997 to 26

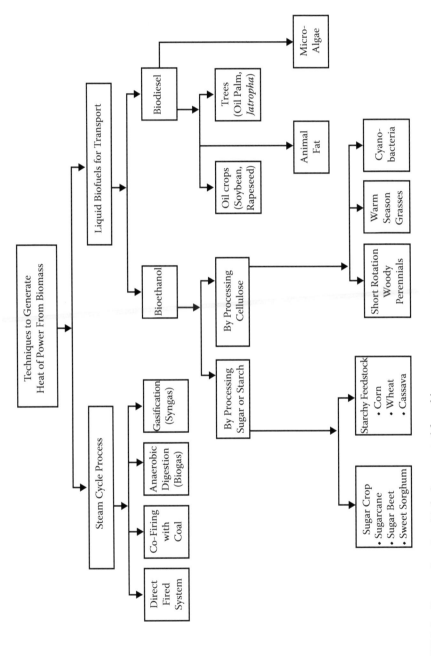

FIGURE 8.2   Range of modern biofuels generated from biomass.

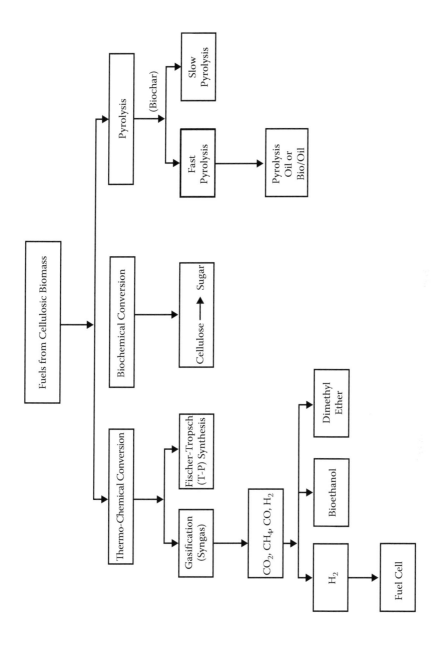

FIGURE 8.3 Techniques for producing fuel from biomass.

BL in 2007 (Table 8.5) and is mandated by the U.S. Congress to be 57 BL by 2015 (Keeney, 2009) and 136 BL by 2022 (Tollefson, 2008). Global biodiesel production has also increased from <1 BL in 1997 to 10 BL in 2007. The use of modern biofuel energy in 2001 was about 6 EJ, that of traditional bioenergy ~39 EJ, compared with the total primary energy consumption of 418 EJ (Smeets and Faiij, 2007; Smeets et al., 2007). The use of bioenergy is equivalent to about 2% of the global NPP and is projected to increase substantially by 2030. In 2007, the biofuel used for transport was 35 BL, compared with 1,200 BL of gasoline use worldwide (Balat, 2007). The use of biofuels as transport energy is expected to grow rapidly. Most countries have a policy to blend 10% of ethanol with petrol for transport fuel. It is in this context that the identification of sources of biomass feedstock for biofuel is extremely important.

CROP RESIDUES AS BIOFUEL FEEDSTOCKS

Crop residues are biomass sources under active consideration (Tillman et al., 2009). Rice is an important source of crop residues (Kim and Dale, 2004), and, because of its low nutritional value as fodder, it is often burned in the field (for example, in Punjab, India) to prepare seedbed for the following wheat crop. Despite numerous benefits of crop residue retention on soil quality (Table 8.6), the sustainability of using crop residues to produce modern biofuels has been widely questioned. The carbohydrate equivalents of crop residues (wheat, rice, corn, barley) are 69% to 76%

TABLE 8.6

Impacts on Soil Quality of Crop Residue Retention

Benefits of Residue Retention	Adverse Impacts of Residue Removal
1. Soil C sequestration to mitigate climate change.	1. Increase in soil erosion and sedimentation.
2. Soil and water conservation.	2. Increase in risk of non-point source pollution.
3. Recycling of plant nutrients.	3. Increase in hypoxia intensity in coastal ecosystems.
4. Habitat and food for soil biota and increase in biodiversity.	4. Adverse impacts on agriculture.
5. Improvement in soil structure and tilth.	5. Damage by flooding to civil structures and waterways.
6. Increase in agronomic production.	6. Soils become main sources of greenhouse gases (CO_2, CH_4, N_2O).
7. Improvements in use efficiency of inputs (e.g., fertilizer).	
8. Necessary component of conservation agriculture and no-till farming.	
9. Suppression of weeds and some pathogens.	
10. Moderation of soil temperature and moisture regimes.	

TABLE 8.7

Estimates of Bioenergy Potential (EJ) of Crop Residues for Major Regions

Year	AFR	CPA	EEU	FSU	LAM	MEA	NAM	PAO	PAS	SAS	WEU	World Total
1990	1.58	3.35	0.75	2.09	1.55	0.48	3.73	0.31	0.61	3.48	1.58	19.51
2050	4.88	5.31	0.92	2.69	4.27	1.45	5.04	0.65	1.55	5.92	2.15	34.83

Source: Recalculated from Fischer and Schrattenholzer, 2001.
AFR = Sub-Saharan Africa
CPA = Centrally Planned Asia and China
CEU = Central and Eastern Europe
FSU = Former Soviet Union
LAM = Latin America and Caribbean
MEA = Middle East and North Africa
NAM = North America
PAO = Pacific OECA
PAS = Other Pacific Asia
SAS = Southern Asia
WEU = Western Europe
World Total = Sum of All Regions

on a dry weight basis and 47% to 50% on a wet weight basis for most cereals, 26% for cassava, 14% for potatoes, and 14% for sugarcane (Aggarwal et al., 2007). The low equivalents of cassava, potato, and sugarcane are due to high moisture content.

Ethanol production potential of most crop residues per tonnes (t) of fresh biomass is 250 L for rice and 265 L for other cereals (Aggarwal et al., 2007). Total energy potential of crop residues (ranging from 5.2 to 17.4 GJ/ha in 1990 and 10.8 to 37.9 GJ/ha in 2050 as estimated by Fischer and Schrattenholzer (2001) was 19.5 EJ in 1990 and is predicted as 34.8 EJ in 2050 (Table 8.7). With energy demand of 365 EJ in 1990, use of crop residues would have provided 3.5% of the global demand. Assuming energy demand of 1000 EJ by 2050, conversion of crop residues into biofuel may meet about 3.5% (34.83 EJ) of the global energy demand (Table 8.7). Other sources of biomass for biofuel, including grasslands, forestlands, and energy plantations, among others, also merit objective and critical evaluations.

The global production of crop residues is estimated at about 4 Gt/year (Lal, 2005). Principal producers of biomass are the U.S., China, and India. China produces some 630 Mt of crop residues per year, of which 37% are directly combusted by farmers (Liu et al., 2008). With the maximum potential of meeting the global energy demand by merely 3.5% through conversion of crop residues into biofuels, there are concerns about the heavy price to be paid for exacerbating risks of soil and environmental degradation and adverse impacts on soil quality and food production.

Despite the benefits of crop residue retention on numerous soil processes, its removal produces several adverse effects on soil quality (Figure 8.4). Principal among adverse impacts are: (1) perturbation in the carbon (C) cycle caused by the negative C budget and depletion of the soil organic matter (SOM) pool, (2) elemental cycling

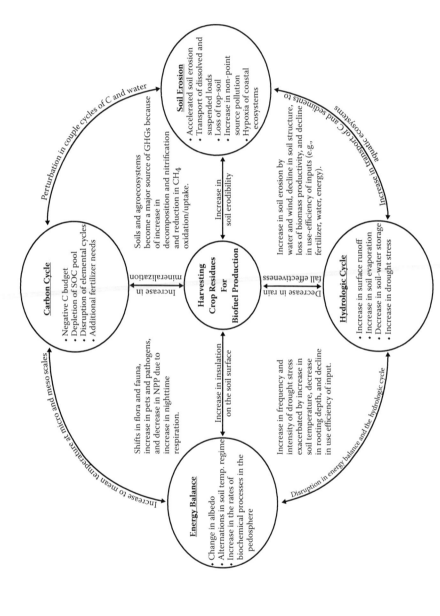

FIGURE 8.4 Adverse effects of crop residue removal on soil quality and agronomic productivity.

due to removal of nutrients contained in crop residues, (3) losses of plant nutrients by acceleration of runoff and erosion on sloping or undulating terrains, (4) effects on hydrologic cycle from increased water losses from surface runoff and evaporation, (5) attendant declines in soil water storage, evapotransportation, and energy balance because of change in surface albedo and alterations in soil temperature regimes, and (6) relative increases in soil chemical and microbial processes such as rate of SOM decomposition (Figure 8.4).

Depending upon the antecedent soil quality, the impact of residue removal on quality may be observed in one to two years in some soils (Blanco-Canqi and Lal, 2007, 2008; Blanco-Canqui et al., 2007; Lal, 2009; Reijnders, 2008; Wilhelm et al., 2004) and after 10 to 15 years or more in others. Nonetheless, the downward spiral of soil degradation and desertification is often set in motion immediately, with long-term adverse impacts on soil quality, productivity, and the environment. In addition, other concerns about using crop residues for biofuel production must be addressed objectively and critically (Figure 8.4). Because of numerous and competing uses of crop residues, it is important to never call these precious commodities "agricultural wastes." They encompass a wide range of materials (crop residues, animal manure, food processing remains, syrups, and plant fibers) and are better described as "agricultural co-products" (Lemke, 2008).

LIFE CYCLE ANALYSES AND NET CARBON GAINS

A complete life cycle analysis (LCA) is needed to assess a net reduction in C emission by substitution of biofuels for fossil fuels. It is argued that biofuels may have a great aggregate environmental cost because of hidden C costs of various processes (including depletion of soil and biotic C pools) than gasoline (Scharlemann and Laurence, 2008; Campbell et al., 2009). It has also been suggested that diversion of corn grains to ethanol production in the U.S. may have accentuated the rate of deforestation of the Amazon (Laurance, 2007) while increasing global food prices (Trostle, 2008) and aggravating food insecurity by decreasing access to low-income populations (Ziegler, 2007).

To be thorough, a complete LCA of corn–soybean agroecosystems for bio-based production must include data on fossil fuel usage, GHG emissions, energy consumption, C, N, P, and major pesticides. It should also incorporate U.S. EPA criteria and pollutants resulting from processes such as fertilizer production, energy production, and on-farm chemical and equipment use (Landis et al., 2007). Crutzen et al. (2008) observed that N_2O release from agro-biofuel production (global warming potential [GWP] of 310) can negate global warming reduction by replacing fossil fuels. Searchinger et al. (2008) reported that use of croplands for biofuel increased GHG emissions.

All C costs must be evaluated, from seeding to harvesting, and chemical conversion to fermentation and distillation, to compute the budget and assess the net C gains in the process. Several factors must be considered in assessing the C neutrality of biofuels (Figure 8.5). An example of a comparative LCA of corn grain production by a commercial farm in the U.S. shows a C output:input ratio of 7:3 for conventional till and 9:1 for no-till operation (Table 8.8). Assuming that no-till farming enhances corn grain production (it may not do so in environments where soil is compacted,

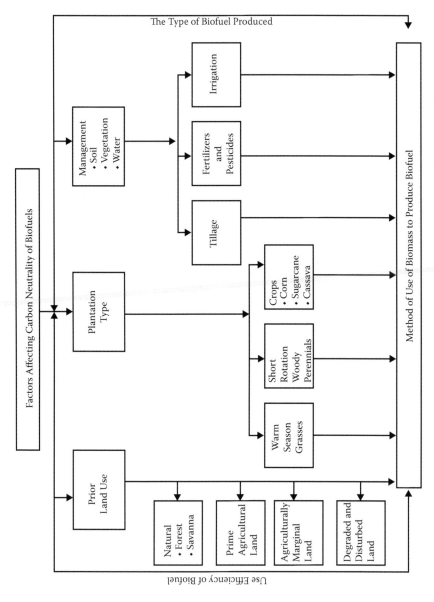

FIGURE 8.5   Life cycle analysis is needed to access C footprints of production systems and C neutralities of biofuels.

TABLE 8.8
Carbon Footprints of Conventional Till and No-Till Corn

Parameter	Conventional Till		No-Till	
	Quantity ha⁻¹	Kg C ha⁻¹	Quantity ha⁻¹	Kg CE ha⁻¹
	Input			
Labor	11.4 hr	44.0	11.4 hr	44.0
Moldboard plow	1	15.2	–	0
Disking	2	16.6	–	0
Harrowing	1	2.0	–	0
Seeding	1	3.2	–	3.8
Machinery	55 kg	97.0	20 kg	35.3
Nitrogen	153 kg	198.9	153 kg	198.9
Phosphorus	65 kg	13.0	65 kg	13.0
Potassium	77 g	11.6	77 kg	11.6
Lime	1,120 kg	179.2	2,000 kg	320.0
Seeds	21 kg	49.6	25 kg	59.0
Herbicides	6.2 kg	39.1	9 kg	56.7
Insecticides	2.8 kg	14.3	4 kg	20.4
Irrigation	8.1 cm	100.0	0	0
Electricity	13.2 kwh	3.2	132 kwh	3.2
Shredding	0	0	1	4.4
Transport	204 kg	16.1	204 kg	16.1
Total		803.0		786.1
	Output			
Grain yield	8,955 kg	2,969.4	9,000 kg	3,087.8
Stover yield	8,655 kg	3,462.0	9,000 kg	3,600.0
Total		6,431.4		6,687.8
Soil erosion (20% emission)	–60		0	
Carbon sequestration in soil O	–500			500
Net Carbon output	5,871.4			7,187.8
C output:input ratio	7:3			9:1

Source: Input and output data from Pimentel, D. and M. Pimentel. 2007. *Food, Energy and Society.* Taylor & Francis, Boca Raton, FL.

Note: If yield reduction with NT is 10%, the output:input ratio is the same as that of conventional tillage.

prone to anaerobiosis because of slow internal drainage, and experiences suboptimal temperatures in spring), C output:input ratio may be higher with no-till or conservation agriculture (CA) systems than with conventional farming. With a corn grain (and stover) reduction of 10% by no-till, the C output:input ratio would be the same for both tillage systems. Long-term experiments in no-till and CA systems in the Midwestern U.S. demonstrated that adoption of best management practices (BMPs) can increase SOM pools in agricultural soils similar to those under woodlands (Olson, 2007). These results suggest that soils can be used for cropland with CA systems and still store C similar to soils in woodlands. Thus, most agricultural soils have large C sink capacities that can be filled through adoption of BMPs.

ENERGY PLANTATIONS

Biomass is defined as organic matter derived from biological organisms (plants, animals, and microorganisms such as green algae). Biomasses generated as co-products of agriculture and forestry include crop residues, wood chips, sawdust, husks, bagasse, nutshells, molasses, food waste, lawn clippings, and municipal waste (Figure 8.6). For any modern biofuel program to be successful, the biofuel feedstock must be produced appropriately and in a sustainable manner. While alternatives to crop residues for sustaining agricultural productivity and natural resources conservation are under consideration (Powell and Unger, 1998), it is also important to identify additional sources of biofuel feedstock.

    Rhetoric aside, biomass for conversion into liquid biofuels must be produced on the basis of sound scientific principles of soil, water, and nutrient management. Above all, a successful biomass feedstock production system must also have a positive soil C budget as indicated by an increasing trend in the SOM pool over time. Thus, it is urgent to identify additional sustainable feedstock sources other than corn (*Zea mays* L.) grains, crop residues, cassava (*Manihoc esculenta*) starches, sugarcane (*Saccharum officinarum*), sweet sorghum (*Sorghum bicolor*), and sugar beet (*Beta vulgaris*). To minimize competition with food, the strategy is to produce cellulosic biomass that can be converted into bioethanol, biogas, and other products (biochar) of economic importance. Biomass from energy plantations can be an effective pathway to low C economy (Anonymous, 2009). Of the 136 BL of ethanol to be produced in the U.S. by 2022, 44% must come from cellulosic feedstock (Tollefson, 2008).

    The conversion of cellulosic biomass to biofuels (bioethanol) is still a work in progress because: (1) the conversion of cellulose and lignin to sugar and starch on an industrial scale is not yet practical (Lynd et al., 2005), (2) identification of genetically modified organisms is required to efficiently ferment a variety of sugars released from conversion of cellulosic biomass, (3) agronomic packages of BMPs must be developed to establish energy plantations involving dedicated species for site-specific soils and climatic conditions, and (4) appropriate research and development should determine steps for economic harvesting, collection, storage and transport of bulk volume of low density materials.

    While scientists are pursuing a cost-effective industrial-scale process to convert cellulose into ethanol (as of 2009 not a single commercial-scale cellulosic ethanol plant is in operation), it is equally important to identify sustainable sources of

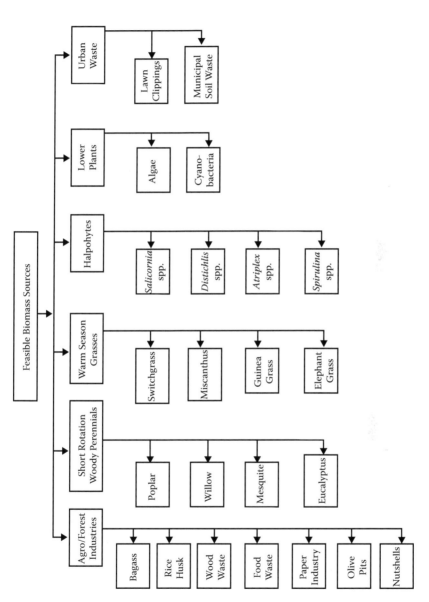

FIGURE 8.6 Feasible sources of biomass (other than crop residues) for heat and power generation.

cellulosic feedstock—for as much as 1 Gt in the U.S. (Somerville, 2006), and 5 Gt or more elsewhere (Ragauskas et al., 2006). Scientific advances in breeding improved varieties and developing BMPs to grow food crops (grain cereals and food legumes) triggered the Green Revolution of the 1960s and saved hundreds of millions from starvation. We now face an urgency to breed cellulosic energy crops and develop agronomic BMPs for adequate and sustained biomass production.

Of course, neither food grains nor residues of agricultural crops should be used for production of biofuels. Three other crops of ligno-cellulosic feedstocks that may be grown on energy plantations of dedicated species are: (1) warm season grasses (WSGs), (2) short-rotation woody perennials (SRWPs), and (3) microalgae and cyanobacteria. Using low intensity and high diversity (LIHD) natural prairie vegetation has also been proposed as a possible source of ligno-cellulosic biomass (Tilman et al., 2006). However, achieving two to three harvests per year to produce 10 t/ ha of dry biomass would be a challenge without additional inputs. Because WSGs and SRWPs are not grown for seeds or fruits, an important selection criterion is the amount of biomass production above ground and below ground. While above-ground biomass is for harvesting as biofuel feedstock, the below-ground biomass is essential to improve the SOM pool and enhance soil quality.

Some examples of WSGs with large biomass productivity are miscant-hus (*Miscanthus giganteus*), switchgrass (*Panicum virgatum* L.), big blue stem (*Andropogan gerardi* Vitnam), Indian grass (*Sorghastrum nutans* L. Nas), blue giant grass (*Calamagrostis canadensis* Michx Bean L.), guinea grass (*Panicum maximum*), elephant grass (*Pennisetum purpureum* Schm), Kallar grass (*Leptochloa fusca*), and others. Switchgrass is native to the U.S. Corn Belt region (e.g., Indiana), and miscanthus is related to sugarcane and is native to East Asia. Miscanthus yields more biomass than switchgrass. Although switchgrass has been widely advertised, its agronomic productivity is low, it is often difficult to establish from seed, and like other monocultures grown for multiple harvests, requires nutrient inputs and main-tenance. Tropical grasses (guinea grass, elephant grass) exhibit high biomass pro-ductivity and require low maintenance, while others can tolerate saline soils (Kallar grass and salt grass or *Distichlis palmeri*). The choices of SRWP species suitable for tree plantations include poplar (*Populus* spp.), willow (*Salix* spp.), mesquite (*Prosopis* spp.), eucalyptus (*Eucalyptus globulus*), and acacia (*Acacia* spp.). Establishment of such plantations must not cause additional deforestation or drainage of peat soils (Fagione et al., 2008; Righelato and Spracklen, 2007).

Varvel et al. (2008) compared corn and switchgrass production on marginal soils for bioenergy in eastern Nebraska. They observed that potential ethanol yield for switch-grass was equal to or greater than the total potential ethanol yield of corn grain and harvested stover fertilized at the same optimum N rate. They concluded that the effects of crop residue removal on agronomic productivity must be studied for major soils and principal ecoregions. Schröder et al. (2008) argued that biotechnology is available to apply fast breeding to promising energy plant species so that valuable land is preserved for agriculture and proposed that opportunities to switch from low-income agriculture to biofuel production may be economically rewarding for small-scale farmers.

Some algae are also suited to producing large quantities of biomass. Appropriate examples include *Betryococcus braunii* and *Spirulina geitleri*. Christi (2007a) concluded that biodiesel from microalgae may be the only renewable biofuel with the potential to completely displace petroleum-derived transport fuel. Average annual productivity of microalgal biomass in a well-designed production system can be 1.53 kg m^{-3} d^{-1}. With average oil content of 30% on a dry mass basis, it would yield oil production at an annual rate of 123 m^3 ha^{-1} of land (Christi, 2007b). Such algae and cyanobacteria can be grown on wastewater that contains plant nutrients.

LAND, WATER, AND NUTRIENT DEMANDS

Similar to food crops, bioenergy plantations of WSGs and SRWPs must be established on good quality lands that do not compete with agricultural production. To meet the goals of biomass production, lands for energy plantations must have high quality soils. Crop production operations and highly productive energy plantations should not be established on soils of marginal quality (too shallow, too steep, too rocky, too dry, too wet) or in too-hot or too-cold climates that yield marginal net primary productivity (NPP). If WSGs such as miscanthus and switchgrass are to be harvested twice or thrice every year, the soils supporting such plantations must be managed to ensure that neither nutrient deficiency nor drought stress affects yield. Rather than the expected annual productivity of 8 to 10 t ha^{-1} of dry biomass, unmanaged switchgrass produces only 1 to 2 t ha^{-1}. Application of N and P can greatly enhance NPPs of such plantations. In addition to N, P, K, and S nutrients, these plantations also require water. Consumptive water use of future biomass energy supply must be carefully assessed (Bernades, 2008).

For energy plantations of dedicated species to be successful, the following requirements are essential: (1) identification of land with high quality soil and water resources; (2) choice of land area that does not compete for crop or livestock production; (3) strong knowledge of BMPs for high biomass production including choice of species, frequency of harvesting, rates of fertilizer application, seeding methods, etc.; and (4) management with high C output:input ratio from seeding to biomass harvest and eventual conversion to liquid biofuel.

Energy plantations can be established on degraded soils—soils that were once biologically productive but have undergone drastic quality declines because of land misuse and soil mismanagement over time, for example, eroded, salinized, compacted, depleted, and polluted soils. Establishment of energy plantations on these soils would restore their quality via a gradual increase in the SOM pool and improvements in soil structure and water retention and transmission properties. However, expectations of biomass production from degraded soils should not be high. Furthermore, high initial costs may be incurred, for example, to restore terrain of eroded soils, provide drainage and install waterways in waterlogged soils, add lime to acidic soils, apply compost and manure in soils depleted of SOM, and apply essential N, P, K, and S nutrients in highly depleted soils. Unless these restorative measures are adopted, stand establishment and rate of growth of plantations established on degraded soils may not be sufficient to provide a viable feedstock source.

RESEARCH AND DEVELOPMENT PRIORITIES

Several relevant issues require research if we are to meet biofuel production targets in the U.S. and globally:

Soil quality—We must evaluate the effects of harvesting crop residues on soil quality. Impacts of residue management on soil quality must be assessed with regard to soil organic carbon dynamics, activity and species diversity of soil fauna (macro, meso, and micro), microbial biomass C, soil structure and susceptibility to crusting and compaction, water retention and transmission properties such as runoff and infiltration, soil water storage and evaporation, efficiency of inputs, and crop yields.

Environmental quality—It is also important to assess the impacts of crop residue removal on environment quality. Special consideration must be given to the non-point source pollution, transport of dissolved and suspended loads in aquatic ecosystems, emissions of GHGs (CO_2, CH_4, N_2O), and assessment of ecosystem C budgets on landscape and watershed bases.

Dedicated species—Identification of appropriate species is required to establish energy plantations for soil-specific conditions within an ecoregion. The choice of species must depend on the ease of stand establishment, adaptation to temperature and moisture regimes, tolerance to pests and pathogens, adaptation to multiple harvests (ratooning) per year, and high NPP and biomass yield.

Best management practices—Similar to BMPs for grain and food crops, BMPs must be developed for energy plantations. They should be soil-specific to ensure sustainability. In addition to production issues, BMPs must also be evaluated in terms of impact on soil quality, water quality, ecosystem C budget (C output:input ratio), and economics of production.

Cellulose conversion—Chemical and biochemical processes to convert biomass into sugar at industrial scale and ferment sugar into ethanol, methanol, and biogas must be developed. Economic feasibility of these processes, an extremely important criterion, must be evaluated.

Land area—Identifying appropriate areas in terms of soil quality and physical accessibility is extremely important. Land area selected for energy plantations must not involve deforestation of tropical rainforests and other ecologically-sensitive ecoregions, drainage of wetlands or peat soils, or cultivation of steep lands. Lands selected must not replace croplands or grazing lands needed for food production.

Energy production and life cycle analysis—The energy contained in biofuels must be critically assessed in relation to the energy consumed to produce them. The energy output:input ratio must be evaluated by conducting detailed life cycle analyses.

CONCLUSIONS

Traditional biofuels (crop residues, animal dung, and fuel wood) have been used as sources of household energy since the dawn of civilization, often with severe adverse impacts on soils, environment, and human health. Modern biofuel production has seen rapid growth, especially since the 1990s. Modern biofuels include bioethanol, biomethanol, biodiesel, and biogas. Bioethanol production from corn grains in the U.S. and from sugarcane in Brazil and biodiesel production from soybeans, oil palm, canola, and crops have caused serious concerns about diverting food grains to bio-fuel production when 1,020 billion food-insecure people worldwide face increases in the prices of food staples.

This has led to strong emphasis on producing cellulosic ethanol and other modern biofuels from cellulosic biomass. Crop residues and other agricultural co-products serve as sources of ligno-cellulosic biomass. However, harvesting crop residues may exert severe adverse impacts on soil quality, water quality, and agronomic production. Retention of crop residues on soil surfaces as mulch provides numerous ecosystem services by enhancing water quality, C sequestration, agronomic yields, and biodi-versity. Thus, indiscriminate removal of crop residues for biofuel production is not a viable option. Energy plantations of dedicated species must be established on good quality soils and on lands that do not compete with food production. Establishment of such plantations on marginal soils and in harsh climates is not a viable option. Restoration of soil quality may be required to establish economically viable planta-tions on degraded or desertified soils. Numerous knowledge caps remain and neces-sitate both basic and applied research to evaluate the potential of site-specific energy plantations and develop industrial level systems to convert biomass into bioethanol, biodiesel, or biogas.

REFERENCES

Aggarwal, P.K. et al. 2007. Fuel ethanol production from Indian agriculture: Opportunities and constraints. *Outlook Agric.* 36: 167–174.

Anonymous. 2009. Pathways to a low-carbon economy. Version 2, Global Greenhouse Gas Abatement Curve. McKinsey & Co.

Bailis, R., M. Ezzati, and D.M. Kammen. 2005. Mortality and greenhouse gas impacts of bio-mass and petroleum energy features in Africa. *Science* 308: 98–103.

Balat, M. 2007. Global biofuel processing and production trends. *Energy Explor. Exploit.* 25: 195–218.

Bernades, G. 2008. Future biomass energy supply: The consumptive water use perspective. *Int. J. Water Res. Dev.* 24: 235–245.

Blanco-Canqui, H. and R. Lal. 2007. Soil and crop response to harvesting corn residue for biofuel production. *Geoderma* 141: 355–362.

Blanco-Canqui, H. and R. Lal. 2008. Stover removal impacts on microscale soil physical prop-erties. *Geoderma* 145: 335–346.

Blanco-Canqui, H. et al. Soil hydraulic properties influenced by stover removal from no-till corn in Ohio. *Soil Tillage Res.* 9: 144–145.

Borlaug, N.E. 2007. Feeding a hungry world. *Science* 318: 359.

Campbell, J.E., D.B. Lobell, and C.B. Field. 2009. Greater transportation energy and GHG offsets from bioelectricity than ethanol. *Science* 324: 1055–1057.

Christi, Y. 2007a. Biodiesel from microalgae beats bioethanol. *Trends Biotechnol.* 26: 126–131.

Christi, Y. 2007b. Biodiesel from microalgae. *Biotechnol. Adv.* 25: 294–306.

Crutzen, P.J. et al. 2008. N_2O release from agro-biofuel production negates global warming reduction by replacing fossil fuels. *Atm. Chem. Phys.* 8: 389–395.

Dasgupta, S. et al. 2009. Improving indoor air quality for poor families: a controlled experi-ment in Bangladesh. *Indoor Air* 19: 22–32.

Demribas, A. 2007. Progress and recent trends in biofuels. *Progr. Energy Combust. Sci.* 33: 1–18.

EIA. 2006. *International Energy Outlook.* Energy Information Administration, U.S. Department of Energy, Washington.

EIA. 2007. *Annual Energy Review.* Energy Information Administration, U.S. Department of Energy, Washington; www.eia.gov/iea.

EPA. 2008. *Inventory of U.S. Greenhouse Gas Emissions and Sinks: 1990–2006.*

FAO. 1009. 1.02 billion people hungry. FAO Newsroom, Rome, Italy.

Fagione, J. et al. 2008. Land clearing and the biofuel carbon debt. *Science* 319: 1235–1238.

Fischer, G. and L. Schrattenholzer. 2001. Global bioenergy potential through 2050. *Biomass and Bioenergy* 20: 151–159.

Goldemberg, J. 2007. Ethanol for a sustainable energy future. *Science* 315: 808.

Goldemberg, J. and P, Guardabassi. 2009. Are biofuels a feasible option? *Energy Policy* 37: 10–14.

Gustafsson, O. et al. 2009. Brown clouds over South Asia: Biomass or fossil fuel combustion? *Science* 323: 495–498.

Habib, G. et al. 2004. New methodology for estimating biofuels consumption for cooking: Atmospheric emissions of black carbon and sulfur dioxide from India. *Global Biogeo. Chem. Cycles* 18: 1–11.

IPCC. 2007a. *Climate Change 2007: The Physical Science Basis.* Cambridge University Press, Cambridge.

IPCC. 2007b. *Climate Change 2007: Mitigation of Climate Change.* Cambridge University Press, Cambridge.

Jacbonson, M.Z. 2002. Control of fossil-fuel particulate black carbon and organic matter, possibly the most effective method of slowing global warming. *J. Geophys. Res.* 16: 1–22.

Keeney, D. 2009. Ethanol USA. *Environ. Sci. Technol.* 43: 8–11.

Kim, S. and B.E. Dale. 2004. Global potential bioethanol production from wasted crops and crop residues. *Biom. Bioenergy* 26: 361–375.

Koonin, S.E. 2008. The challenge of CO_2 stabilization. *Elements* 4: 294–295.

Kumie, A. et al. 2009. Magnitude of indoor N_2O from biomass fuels in rural settings of Ethiopia. *Indoor Air* 19: 14–21.

Lal, R. 2005. World crop residues production and implications of its use as biofuel. *Environ. Int.* 31: 575–584.

Lal, R. 2009. Soil quality impacts of reside removal for bioethanol production. *Soil Tillage Res.* 102: 233–241.

Landis, A.E., S.A. Miller, and T.L. Theis. 2007. Life cycle of the corn–soybean agroecosystem for bio-based production. *Environ. Sci. Technol.* 41: 1457–1464.

Laurence, W.F. 2007. Switch to corn promotes Amazon deforestation. *Science* 318: 1721.

Lemke, D. 2008. Redefining ag wastes as co-products. *Biores. Technol.* 99: 5250–5260.

Liu, H. et al. 2008. Distribution, utilization structure and potential of biomass resources in rural China: With special reference to crop residues. *Renew. Sustain Energy Rev.* 12: 1402–1418.

Lynd, L.R. et al. 2005. Consolidated bioprocessing of cellulosic biomass: Update. *Curr. Opin. Biotechnol.* 16: 577–583.

Mathews, J.A. 2007. Biofuels: What a biopact between North and South could achieve. *Energy Policy* 35: 3550–3570.

Mathur, J., N.K. Bansal, and H.J. Wagner. 2004. Dynamic energy analysis to assess maximum growth rates in developing power generation capacity: Case study of India. *Energy Policy* 32: 281–287.

Mirza, U.K., N. Ahmad, and T. Majeed. 2008. An overview of biomass energy utilization in Pakistan. *Renew. Sustain. Energy Rev.* 12: 1988–1996.

Mudway, I.S. et al. 2005. Combustion of dried animal dung as biofuel results in the generation of highly redox active fine particulates. *Particle Fibre Toxicol.* 2: 1–11.

Muir, J.P. 2001. Dairy compost, variety and stand age effects on Kenaf forage yield, nitrogen and phosphorus concentration and uptake. *Agron. J.* 93: 1169–1173.

Naidu, B.S.K. 1996. Indian scenario of renewable energy for sustainable development. *Energy Policy* 24: 575–581.

Naylor, R.L. et al. 2007. The ripple effect: Biofuels, food security and the environment. *Environment* 49: 30–43.

Ohlrogge, J., D. Allen, B. Berguson, D.D. Penna, Y. Shachar-Hill, and S. Stymne. 2009. Driving on biomass. *Science* 324: 1019–1020.

Olson, K.R. 2007. Soil organic carbon storage in southern Illinois woodland and cropland. *Soil Sci.* 172: 623–630.

Pachauri, S. 2004. An analysis of cross-sectional variations in total household energy require-ments in India using micro-survey data. *Energy Policy* 32: 1723–1735.

Pachauri, S. and D. Spreng. 2002. Direct and indirect energy requirements of households in India. *Energy Policy* 30: 511–523.

Pimentel, D. and M. Pimentel. 2007. *Food, Energy and Society*. Taylor & Francis, Boca Raton, FL.

Powell, J.M. and P.W. Unger. 1998. Alternative to crop residues for sustaining agricultural productivity and natural resources conservation. *J. Sustainable Agric.* 11: 59–84.

Proquest Historical News. 1925. Ford predicts fuel from vegetation. *New York Times*, Sept. 24, p. 24.

Ragauskas, A.J. et al. 2006. The path forward for biofuels and biomaterials. *Science* 311: 484–489.

Ramanathan, V. et al. 2001. Aerosols, climate and the hydrological cycle. *Science* 294: 2119–2124.

Ravindranath, N.H. 1997. Energy options for cooking in India. *Energy Policy* 25: 63–75.

Reddy, B.S. 2003. Overcoming the energy efficiency gap in India's household sector. *Energy Policy* 31: 1117–1127.

Regalbuto, J.R. 2009. Cellulosic biofuels—got gasoline? *Science* 325: 822–824.

Reijnders, L. 2008. Ethanol production from crop residues and soil organic carbon. *Res. Conserv. Policy* 52: 653–658.

Righelato, R. and D.V. Spracklen. 2007. Carbon mitigation by biofuels or by saving and restor-ing forests. *Science* 317: 902.

Scharlemann, J.P.W. and W.F. Laurence. 2008. How green are biofuels? *Science* 139: 43–44.

Schröder, P. et al. 2008. Bioenergy to save the world. *Env. Sci. Pollut. Res.* 15: 196–204.

Searchinger, T. et al. 2008. Use of U.S. croplands for biofuel increases greenhouse gases through emissions from land use change. *Science* 319: 1238–1240.

Singh, G. et al. 1992. Soil erosion rates in India. *J. Soil Water Conserv.* 47: 97–99.

Smeets, E.M.W. and A.P.C. Faiij. 2007. Bioenergy potentials from forestry in 2050: An assess-ment of drivers that determine the potentials. *Climate Change* 81: 353–390.

Smeets, E.M.W. et al. 2007. A bottom-up assessment and review of global bioenergy poten-tials to 2050. *Progr. Energy Combust. Sci.* 33: 56–106.

Somerville, C. 2006. The billion ton biofuel vision. *Science* 312: 1277.

Srivastava, L. 1997. Energy and CO_2 emissions in India: Increasing trends and alarming por-tents. *Energy Policy* 25: 941–949.

Sticklen, M.B. 2007. Feedstock crop genetic engineering for alcohol fuels. *Crop Sci.* 47: 2238–2248.

Suganthi, L. and A. Williams. 2000. Renewable energy in India: A modeling study for 2020–2021. *Energy Policy* 28: 1095–1109.

Suresh, D.S. et al. 1998. Siltation analysis in the Neyyar reservoir and forest degradation in its catchment: A study from Kerala State, India. *Env. Geol.* 39: 390–397.

Tillman, D., J. Hill, and C. Lehman. 2006. Carbon-negative biofuels from low-input high diversity grassland biomass. *Science* 314: 1598–1600.

Tillman, D. R. Socolow, J.A. Foley, J. Hill, E. Larson, L. Lynd, S. Pacala, J. Reilly et al. 2009. Beneficial biofuels—The food, energy and environment trilemma. *Science* 325: 270–271.

Tollefson, J. 2008. Not your father's biofuels. *Nature* 451: 880–883.

Trostle, R. 2008. Global agricultural supply and demand: Factors contributing to the recent increase in food security in food commodity prices. USDA-ERS, Outlook Report (www. ers.usda.gov).

Varvel, G.E. et al. 2008. Comparison of corn and switchgrass on marginal soils for bioenergy. *Biom. Bioenergy* 32: 18–21.

Vasudevan, P.T. and M. Briggs. 2008. Biodiesel production: Current state of the art and challenges. *J. Ind. Microbiol. Biotechnol.* 35: 421–430.

Venkatraman, C. et al. 2005. Residential biofuels in South Asia: Carbonaceous aerosol emissions and climate impacts. *Science* 307: 1454–1456.

WEO. 2002. Energy and Poverty. International Energy Agency. (http://www.iea.org/textbase/nppdf/free/2002/energy_poverty.pdf).

Wild, A. 2003. *Soils, Land and Food: Managing the Land during the 21st Century.* Cambridge University Press, Cambridge.

Wilhelm, W.W. et al. 2004. Crop and soil productivity response to residue removal: A literature review. *Agron, J.* 96: 1–17.

Williams, P.A. 2009. The biofuels landscape through the lens of industrial chemistry. *Science* 325: 707–708.

Yevich, R. and J.A. Logan. 2003. An assessment of biofuel use and burning of agricultural waste in the developing world. *Global Biogeochem. Cycles* 17: 6-1–6.21.

Zhang, J., and K.R. Smith 2005. Indoor air pollution from household fuel combustion in China: A review. Tenth International Conference on Air Quality and Climate, Beijing.

Ziegler, J. 2007. Draft of Special Report on the Right to Food to the United Nations General Assembly, New York.

Index